T0173633

Collins

Exploring Science | Grade 7

Derek McMonagle

Reviewers: Marlene Grey-Tomlinson, Bernadette Ranglin, Maxine McFarlane & Monacia Williams

Collins

William Collins' dream of knowledge for all began with the publication of his first book in 1819. A self-educated mill worker, he not only enriched millions of lives, but also founded a flourishing publishing house. Today, staying true to this spirit, Collins books are packed with inspiration, innovation and practical expertise. They place you at the centre of a world of possibility and give you exactly what you need to explore it.

Collins. Freedom to teach.

Published by Collins
An imprint of HarperCollins*Publishers*
The News Building
1 London Bridge Street
London
SE1 9GF

HarperCollins*Publishers*
1st Floor, Watermarque Building,
Ringsend Road
Dublin 4, Ireland

Browse the complete Collins Caribbean catalogue at
www.collins.co.uk/caribbeanschools

10 9 8 7 6 5

ISBN 978-0-00-826327-0

British Library Cataloguing in Publication Data
A catalogue record for this publication is available from the British Library.

Publisher: Elaine Higgleton
Commissioning editor: Tom Hardy
In-house senior editor: Julianna Dunn
Author: Derek McMonagle
Reviewers: Marlene Grey-Tomlinson, Bernadette Ranglin, Maxine McFarlane & Monacia Williams
Project Manager: Alissa McWhinnie, QBS Learning
Copyeditor: Mitch Fitton
Proofreader: David Hemsley
Photo researcher, illustrator & typesetter: QBS Learning
Cover designer: Gordon MacGilp
Series Designer: Kevin Robbins
Cover photo: Rainer Albiez/Shutterstock
Production controller: Tina Paul
Printed and bound by: Ashford Colour Press Ltd.

MIX
Paper from
responsible sources
FSC™ C007454

FSC
www.fsc.org

This book is produced from independently certified FSC™ paper to ensure responsible forest management.

For more information visit: www.harpercollins.co.uk/green

Contents

Introduction – How to use this book

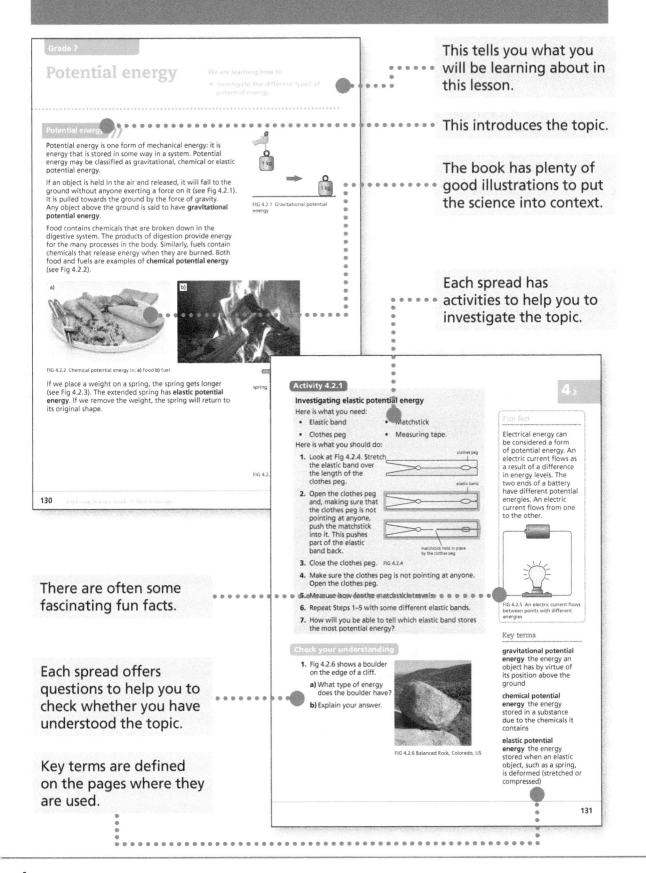

Grade 7

Potential energy

We are learning how to:
• investigate the different types of potential energy.

Potential energy

Potential energy is one form of mechanical energy: it is energy that is stored in some way in a system. Potential energy may be classified as gravitational, chemical or elastic potential energy.

If an object is held in the air and released, it will fall to the ground without anyone exerting a force on it (see Fig 4.2.1). It is pulled towards the ground by the force of gravity. Any object above the ground is said to have **gravitational potential energy**.

Food contains chemicals that are broken down in the digestive system. The products of digestion provide energy for the many processes in the body. Similarly, fuels contain chemicals that release energy when they are burned. Both food and fuels are examples of **chemical potential energy** (see Fig 4.2.2).

FIG 4.2.1 Gravitational potential energy

FIG 4.2.2 Chemical potential energy in: a) food b) fuel

If we place a weight on a spring, the spring gets longer (see Fig 4.2.3). The extended spring has **elastic potential energy**. If we remove the weight, the spring will return to its original shape.

FIG 4.2.

130 Exploring Science Grade 7, Unit 4: Energy

This tells you what you will be learning about in this lesson.

This introduces the topic.

The book has plenty of good illustrations to put the science into context.

Each spread has activities to help you to investigate the topic.

Activity 4.2.1

Investigating elastic potential energy

Here is what you need:
• Elastic band   • Matchstick
• Clothes peg   • Measuring tape.

Here is what you should do:

1. Look at Fig 4.2.4. Stretch the elastic band over the length of the clothes peg.

2. Open the clothes peg and, making sure that the clothes peg is not pointing at anyone, push the matchstick into it. This pushes part of the elastic band back.

3. Close the clothes peg. FIG 4.2.4

4. Make sure the clothes peg is not pointing at anyone. Open the clothes peg.

5. Measure how far the matchstick travels.

6. Repeat Steps 1–5 with some different elastic bands.

7. How will you be able to tell which elastic band stores the most potential energy?

clothes peg

elastic band

matchstick held in place by the clothes peg

Check your understanding

1. Fig 4.2.6 shows a boulder on the edge of a cliff.

   a) What type of energy does the boulder have?

   b) Explain your answer.

FIG 4.2.6 Balanced Rock, Colorado, US

4.2

Fun fact

Electrical energy can be considered a form of potential energy. An electric current flows as a result of a difference in energy levels. The two ends of a battery have different potential energies. An electric current flows from one to the other.

FIG 4.2.5 An electric current flows between points with different energies

Key terms

gravitational potential energy the energy an object has by virtue of its position above the ground

chemical potential energy the energy stored in a substance due to the chemicals it contains

elastic potential energy the energy stored when an elastic object, such as a spring, is deformed (stretched or compressed)

131

There are often some fascinating fun facts.

Each spread offers questions to help you to check whether you have understood the topic.

Key terms are defined on the pages where they are used.

4

Review of Energy

- Energy is the ability to do work.
- Energy in different contexts is described as different forms of energy.
- Forms of energy include: heat, light, sound, electrical energy, chemical energy, nuclear energy, potential energy and kinetic energy.
- Potential energy is energy that is stored in some way and includes: gravitational potential energy, chemical potential energy and elastic potential energy.
- Kinetic energy is the energy that objects have when they move.
- In a swinging pendulum, energy is converted between gravitational potential energy and kinetic energy.
- Nuclear energy is the result of changes to the structure of atoms. It is the source of energy in the Sun and in nuclear power stations.
- Non-renewable energy sources are those that are not replaced by nature at the same rate as they are used up. They include coal, crude oil and natural gas.
- Renewable energy sources are those that are replaced by nature at the same rate as they are used up. They include hydroelectric energy (flowing water), solar energy, geothermal energy, tidal energy and wind energy.
- Biofuels are the result of energy stored by photosynthesis either directly, for example in wood obtained from trees, or indirectly, for example by animals eating plants and producing dung. Biofuels include biogas, wood, charcoal and ethanol.
- Energy can be transformed from one form into others. When this happens, work is done.
- Energy transformations can be shown as Sankey diagrams.
- The world is moving towards a greater use of renewable sources of energy as reserves of non-renewable energy sources are used up.
- Jamaica is currently developing renewable energy resources including solar power, wind power and hydroelectric power.
- In 2009 Jamaica adopted a National Energy Policy which aims to provide 30% of the country's energy needs from renewable sources by 2030.
- Individual people can make a significant contribution to reducing a country's demand for energy.
- All of the Caribbean countries are looking for ways of reducing their reliance on fossil fuels by developing the renewable energy sources which are available to them.

At the end of each group of units there are pages which list the key topics covered in the units. These will be useful for revision.

At the end of each section there are special questions to help you and your teacher review your knowledge, see if you can apply this knowledge to new sitauations and if you can use the science skills that you have developed.

Science, Technology, Engineering, Arts and Mathematics (STEAM) activities are included, which present real-life problems to be investigated and resolved using your science and technology skills. These pages are called **Science in practice**.

Review questions on Energy

1. **a)** What is the difference between potential energy and kinetic energy?
   b) Draw a diagram of a pendulum to show the position of the bob when:
   i) the potential energy is maximum
   ii) the kinetic energy is maximum

2. State one form of energy that can be sensed by the:
   a) eyes   **b)** ears   **c)** skin.

3. **a)** Name the process by which energy is released in a nuclear power station.
   b) Explain how the release of energy from a nuclear fuel like uranium is different than from a chemical fuel like wood.

4. Table 4.RQ.1 shows some of the world's energy sources. They are arranged in order of the total amount of energy they provide.

Energy source	Relative amount of energy provided
Oil	Most energy supplied
Natural gas	
Coal	
Nuclear	↓
Hydroelectric	
Wind farms	Least energy supplied

TABLE 4.RQ.1

   a) From this table give:
   i) a non-renewable source of energy
   ii) a renewable source of energy.
   b) What is meant by a 'non-renewable energy source'?
   c) Name one other large-scale source of energy not given in the table.
   d) If a similar table was drawn up 100 years from now, suggest one way in which it would be different.

Propagating plants from cuttings

Mr Patterson is planning to start a business selling flowering plants. His plan is to produce large numbers of plants for sale in plant centres.

He understands about obtaining plants by growing from seeds but he also wants to propagate plants from cuttings and he is uncertain about the best way to go about this.

FIG 5.SIP.1 A plant centre

Mr Patterson carried out some research at local gardening shops but this made him even more confused. He found that there are a number of products on the market to help him root his cuttings. They all claim to help cuttings to develop roots but which one is best? Will they all produce the same result, in which case he might as well buy the cheapest, or is one product better than the others, in which case it might be worth spending a little more?

FIG 5.SIP.2 Rooting hormone

Mr Patterson is investing a lot of money into his business so it is important that he finds out which of the rooting products is going to give him the best results. Before he starts taking cuttings on a large scale he has hired you to investigate and make recommendations.

1. You are going to work in groups of 3 or 4 to investigate which rooting hormone product works best on plant cuttings. The tasks are:
   - To review how to propagate a plant by taking a cutting.
   - To devise a standard method of taking cuttings.
   - To carry out research into what products which promote cuttings to grow roots are available in your local gardening shops.
   - To make a note of the active chemical(s) in each product and read the instructions for how the product should be used.
   - To plan a fair test in which you will compare the performance of each product.
   - To decide how you will evaluate the results of your test in order to make recommendations.
   - To compile a report, including a PowerPoint presentation in which you explain how you carried out your fair test, show photographs of the results and present the conclusions you have come to about the different products tested.

Unit 1: Science and scientific processes

We are learning how to:

- define science and technology.

Science and technology ≫

What is science?

Science is the study of nature and the environment. The word 'science' comes from the Latin *scientia*, meaning 'knowledge'. In science, **observations** and **experiments** are used to describe and explain natural occurrences. The knowledge gained is never complete, but is always being added to through further research and experiments.

Although the various branches of science are categorised as either social sciences or natural sciences, all your activities and research will be confined to areas in the natural sciences.

FIG 1.1.1 Scientist at work

Activity 1.1.1

Finding out about different sciences

Here is what you should do:

1. Fig 1.1.2 identifies some natural sciences.

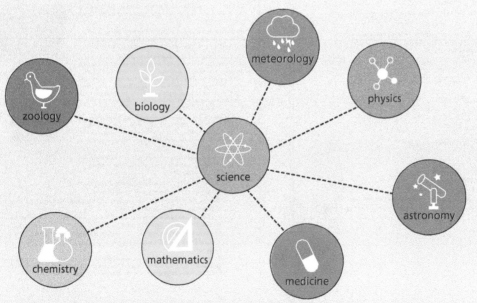

FIG 1.1.2 Some branches of science

In a group, discuss what is studied in each one of the sciences.

2. Match each branch of science with one area of study from the list below:

- Healing
- Living things
- Objects beyond the Earth
- Quantity
- Animals
- The atmosphere
- Behaviour of matter
- Structure of matter.

What is technology?

The word '**technology**' comes from the Greek word *techne*, meaning 'art' or 'skill'. Technology is the **application** of knowledge gained from science in practical ways – to solve problems and improve the quality of life. Using technology, people are able to control and adapt to their natural environments. Our development and use of tools, the building of ships, and the invention of medical devices such as X-ray and MRI scanners are all examples of technology.

How does the scientific process work?

Science is not just about collecting facts to describe the natural world and its origin. It seeks to understand and then provide models for how the world works. Despite its usefulness, science has numerous limitations. For example, although science can provide guidelines to help in the recovery from the passage of a hurricane, it is unable to stop the action of a hurricane. These limitations are used as stepping stones for acquiring even more knowledge with which to address new issues – an example is the creation of hurricane braces to attach roofs more securely to houses.

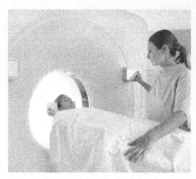

FIG 1.1.3 Doctor using MRI scanner

Fun fact

The first wristwatch was made by Patek Philippe in 1868.

Key terms

science system for studying the physical and living worlds based on experiments

observation collecting information via the senses

experiment the process of conducting a scientific test

technology the use of knowledge gained from science experiments to improve the quality of life

application putting scientific information to special use

Check your understanding

1. Explain the difference between science and technology. Give some examples of scientific discovery and of technological developments.

2. Find out what made each of the following scientists famous:

- Thomas Edison
- Richard Leakey
- Isaac Newton
- Jane Goodall
- Marie Curie
- John Dalton
- Maria Goeppert-Mayer
- Albert Einstein
- Louis Pasteur.

a) b)

FIG 1.1.4 Famous scientists
a) Maria Goeppert-Mayer and
b) Albert Einstein

Impacts of science and technology

We are learning how to:

• make use of science and technology to solve community-based issues.

Impact of science and technology on daily life

Science and technology influence life today in a huge variety of ways. Imagine your day:

- You are woken at 6 a.m. by the music from your clock radio. You jump down from your bunk bed and into your bedroom slippers. You rush to the bathroom, turn on the light and the heater and have a shower. You iron your shirt, get dressed and comb your hair while looking in the mirror.

- Then you go to the kitchen, grab the remote control and turn on the television. You switch on the kettle to make tea and put bread in the toaster. You go to the fridge and find the ingredients for your packed lunch.

- After breakfast, you clean up the kitchen, and put out the rubbish as the garbage truck passes.

All of these aspects of life are influenced by science and technology, which have also contributed to the development of motor vehicles, telephone lines, lifts, space rockets and water-treatment plants and, on a smaller scale, items like washers for tap heads, aglets (metal or plastic tube fixed around each end of a shoelace) and paperclips.

FIG 1.2.1 Some of our daily activities

Group work for class presentation

Here is what you should do:

1. Interview an elderly person and find out how life was for them growing up: the type of recreation activities there were, food, houses, communication, etc. Compare their lifestyle with that of today.

2. Compare how telecommunication services worked and how they have been developed to what they are today.

3. Compare the transport service, from horseback to space travel.

4. Consider the invention of gadgets (for example, a bottle opener) to make work easier.

Despite all these advantages of scientific and technological progress, there can be disadvantages. The demand for paper, wood and rubber reduces the number of trees. Fuels for cars and factories cause pollution. Medication can be accompanied by side effects. The important thing to note here is that technological solutions to scientific problems must be applied carefully and with the proper **precautions**.

Science and technology are also applied widely in medicine, agriculture, defence, economics, leisure and exploration. It does not seem that there is any barrier to the current rate of technological advancement.

Check your understanding

1. List two advantages and two disadvantages of the smartphone.

2. List two advantages and two disadvantages of modern cars.

Fun fact

FIG 1.2.2

In 1968, the first computer mouse was introduced by Douglas Engelbart at a Computer Expo in the USA.

Key term

precaution safety measures

Working like a scientist

We are learning how to:

• interpret safety symbols.

Safety symbols 〉〉〉

There are many safety **hazards** in a science laboratory ('lab'), including many types of chemical. Different chemicals pose different risks. There are internationally recognised symbols designed specifically for lab safety, especially when handling chemicals.

Activity 1.3.1

Looking at safety symbols found on chemicals

1. Fig 1.3.1 shows the seven common safety symbols. Can you match them with what they represent from this list of words?

1

2

3

| toxic | oxidising | radioactive | corrosive |
| explosive | flammable | irritant/harmful | |

4

5

6

7

FIG 1.3.1 Safety symbols

2. Your teacher will show you bottles and containers with symbols. Test your memory by identifying the symbols on them.

3. Draw the safety symbols in your lab book/note book or download and print them. Then write what each one means.

What would you do?

Here is what you should do:

- In a group, role-play a scenario of what could happen if someone comes in contact with a chemical after ignoring its safety symbol, and what should be done.

FIG 1.3.2 Safety symbols are provided on domestic cleaning agents

Check your understanding

1. Check the labels of various products at home and identify the safety symbols. Identify which component of each product most influences the care that should be taken when using it.

2. Carry out research and explain what would happen if someone mixed a product which contains chlorine with another which contains ammonia.

1.3

Key terms

hazard an unavoidable danger or risk

toxic poisonous

oxidising providing oxygen so that substances can burn or react in other ways

corrosive destroying substances by wearing them away while chemically reacting with them

flammable ignites easily

irritant substance that is neither corrosive nor toxic but reacts negatively on contact

Safety in the school laboratory

We are learning how to:

- behave properly whenever we are in the lab
- carry out lab activities
- work in an orderly and careful way
- react if accidents occur.

Working safely 〉〉〉

The **laboratory** is the special place where scientists carry out their work. It is usually referred to as the 'lab'. As budding scientists, it is necessary that you know the rules of conduct that govern all lab activity so that you are always safe when carrying out activities.

Activity 1.4.1

Recognising unsafe activities

Look carefully at the different activities that are taking place in the picture. Make a list of the activities that are unsafe.

FIG 1.4.1

General lab rules

These general rules should be followed when working in the lab:

1. Permission is needed to enter the lab.

2. No food or drink should be brought into the lab.

3. Open all doors and windows unless otherwise directed.

4. Always walk in the lab. Do not run or play around.

5. Store all coats, bags and other belongings tidily.

6. Do not tamper with any electrical mains or fittings.

7. Do not attempt experiments without the teacher's permission.

8. Read all instructions carefully and ask for clarification as required.

9. Handle apparatus and materials both carefully and correctly.

10. Report immediately any broken or damaged apparatus.

11. Do not remove apparatus or chemicals from the lab.

12. Work tidily, clean all apparatus thoroughly and dispose of waste correctly.

13. Do not hold hot apparatus with your bare hands.

14. Wash hands thoroughly before leaving the lab or experimenting.

When accidents occur:

- Report all accidents, injuries, breakages or spillages to your teacher immediately.

- If a chemical gets into your mouth, spit it out and rinse your mouth as quickly as possible with plenty of water and report the incident to the teacher.

- If any chemical falls on your body or clothing, rinse the area thoroughly with water and report the incident to the teacher.

Check your understanding

1. Say why it is necessary to have lab rules.

2. How many of the laboratory rules listed were not being observed in the picture on page 12?

3. Why should you not pour unused chemicals back into the bottles?

4. List some accidents that may occur in the lab.

Key term
....

laboratory a room used for scientific experiments

Famous scientists

Famous scientists >>>

Over the centuries that science has evolved, there have been many important contributions by individuals but, as in many areas or life, some are remembered while others have long been forgotten.

It is often, but not always, the case that we remember those who made the most significant contributions.

Some scientists remain famous because they have **laws** named after them. Later in your course you will learn about Newton's Laws of Motion.

Isaak Newton.

FIG 1.5.1 Sir Isaac Newton

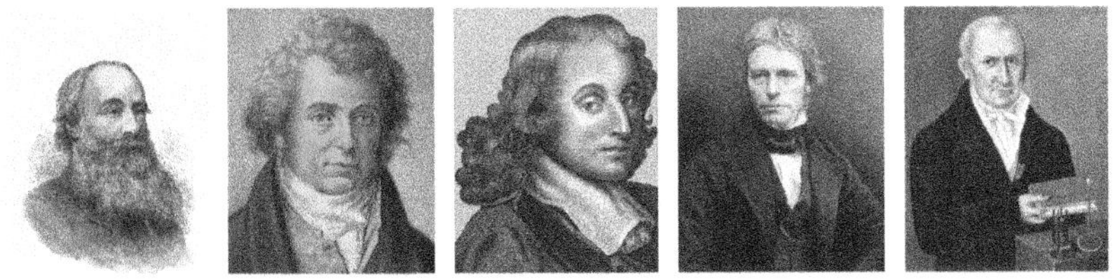

| James Joule | Andre-Marie Ampere | Blaise Pascal | Michael Faraday | Alessandro Volta |

FIG 1.5.2 Scientists who have given their names to units

Some scientists have been honoured by having a **unit** named after them. The newton is the unit of force. Some other examples are given in Fig 1.5.2. What quantity is measured in each of the units named after these scientists?

Some scientists have been honoured by having an **element** named after them. These include:

Bohrium (Niels Bohr)

Copernicium (Nicolaus Copernicus)

Curium (Pierre and Marie Curie)

Einsteinium (Albert Einstein)

Some scientists have given their names to **apparatus** which are still used today.

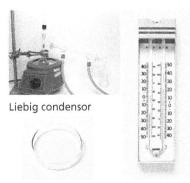

Liebig condensor

Petri dish   Six's thermometer

FIG 1.5.3 Apparatus named after the inventor

Contributions made by Jamaican scientists

The following eight Jamaican scientists each made a significant contribution to science during the last century.

Harold M. Johnson   G. Lalor

Cicely Williams   Kenneth Richards

William E. McCulloch   Paula Tennant

Leigh D. Lord   Thomas P. Lecky

Here is what you should do:

1. Choose one of these scientists who will be the centre of your study.

2. Carry out research into the contribution made by this scientist. You should investigate things like:

   a) In what branch of science did they work?

   b) What was it that they discovered or did?

   c) How did this change peoples' lives for the better?

3. Write a brief report on the work of this scientist that you could read out to the class if invited by your teacher.

Check your understanding

1. Suggest the name of the scientist who invented the following apparatus.

   a) Newtonian telescope   b) Torricellian barometer

   c) Galilean thermometer   d) Bunsen burner

2. Each of the following units is named after a scientist.

   watt   celsius   hertz   ampere

   Which unit is a measure of:

   a) temperature?   b) electrical current?

   c) power?   d) frequency?

Key terms

law statement based on repeated experimental observations that describes some aspects of the universe

unit something used to express an amount of a physical quantity

element substance that cannot be broken down into simpler substances

apparatus instruments and containers used to carry out experiments

Scientific skills

We are learning how to:

- use scientific skills to carry out an experiment using the scientific method.

The scientific method 〉〉

Scientists carry out their work or research in an organised, or systematic, way called the **scientific method**. Why is it necessary for scientists to be so organised?

To become a scientist, you need to develop a number of skills to enable you to carry out research that gives accurate results. Inaccurate results give a false picture of the subject under research. It is therefore important for scientists to be very careful and organised.

The table below gives five scientific skills with their definitions. You will develop these skills as you work through this course.

FIG 1.6.1 Agricultural scientist at work

Skill	Definition
Hypothesising	Suggesting a possible explanation for things in a way that can be verified
Planning and conducting experiments	Carrying out a process to answer a specific question
Collecting data	Gathering numerical information
Recording and reporting	Ensuring data is written down securely and then **communicating** it in tables, diagrams or charts so that it is easily understood
Analysing data	Looking for patterns and information within sets of data

TABLE 1.6.1 Scientific skills

Scientists use all the skills listed in Table 1.6.1 when they conduct experiments.

Experimenting scientifically

You will carry out an experiment on piercing balloons.

Scientists always have a reason for carrying out an experiment. Now think of all what you will do and why – suppose you place sticky tape on the balloon and pierce through the tape, what **prediction** can you make?

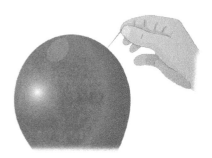

FIG 1.6.2

When the experiment is completed you will write a report, so you must be extremely observant during the experiment.

Before conducting an experiment, it is important that you check that:

- everything you need to use is available and close at hand
- all instructions are carefully read and understood.

Activity 1.6.1

Here is what you need:
- Balloons
- Clear sticky tape
- A straight pin
- Scissors.

Here is what you should do:

1. Inflate two balloons fully and tie each one at the neck.

2. Pierce one of the balloons with the pin. Do not let go of the pin. Carefully observe what happens.

3. Place sticky tape on the other balloon so that there are no air pockets between the tape and the balloon.

4. Pierce this balloon through the tape. Do not let go of the pin. Carefully observe what happens.

5. Record what you observed with both balloons.

6. Provide suggestions to explain what you observed.

Check your understanding

1. Why did you have to hold on to the pin?

2. Why was it important that there were no air pockets between the tape and the balloon?

3. What caused the first balloon to burst so quickly?

4. Was the behaviour of the air in both balloons the same or different?

5. Give a reason to support your answer.

Fun fact

The microwave oven was invented by a researcher who, when he walked past a radar tube, found that the chocolate bar in his pocket melted.

Key terms

scientific method the systematic manner of conducting experiments

communicating translating information into tables, diagrams and words for them to be easily understood

planning formulating an orderly set of events which may lead to achieving a goal

hypothesising suggesting a possible explanation for things in a way that they can be verified

analysing examining something in detail in order to discover meaning, patterns, essential features, etc.

predicting stating expected outcomes based on experience

Carrying out a fair test

We are learning how to:

- identify things that may vary in an experiment
- carry out a fair test.

Variables

A **constant** is something that has a fixed value. The number of days in a week is a constant; the value is always 7.

A **variable** is something that can take a range of values. The amount of money you have in your pocket is a variable. Sometimes you might have lots of dollars and other times you might have nothing.

You will learn more about different kinds of variables in Grade 8. In this lesson you are going to focus on identifying variables in an experiment. You will learn why they are important in the context of carrying out a fair test or comparison.

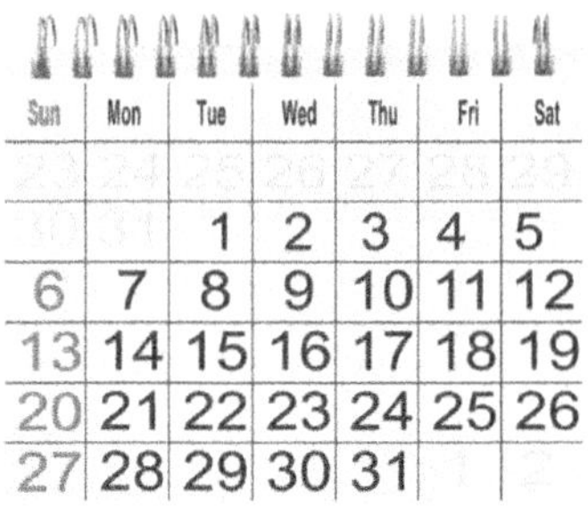

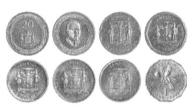

FIG 1.7.1 The number of days in a week is a constant

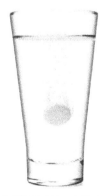

FIG 1.7.2 The amount of money in your pocket is a variable

Identifying the variables in an experiment

A student carried out an experiment to compare how quickly three substances dissolve in water.

For each substance, he added some of the solid to water and timed how long to took to dissolve. His results are summarised in the following table.

Substance	Mass used / g	Volume of water / cm^3	Temperature of water / °C	Stirring	Time taken to dissolve / min
A	1.03	25	25	No	7
B	0.85	30	32	Yes	2
C	0.92	20	19	Yes	5

TABLE 1.7.1

On the basis of his results the student decided that Substance B dissolves quickest and Substance A dissolves slowest. Use these statements to decide if this was a **fair test**.

- The greater the mass of a substance, the longer it will take to dissolve.
- The greater the volume of water, the quicker a substance will dissolve.
- In general, the warmer the water, the quicker a substance will dissolve.
- Stirring helps substances to dissolve more quickly

FIG 1.7.3 Dissolving substances in water

It might be that Substance B only dissolved quickest because, compared to the other substances:

- there was a smaller mass of it
- it was placed in a larger volume of water
- it was placed in warmer water
- it was stirred.

The mass of the substance, the volume of water, the temperature of the water and whether the mixture was stirred are all variables. We can only be sure the results are valid if these variables are kept the same for all three substances.

Activity 1.7.1

Comparing how quickly three substances dissolve

Here is what you need:

- Beaker 100 cm³ × 3
- Substances × 3
- Balance
- Measuring cylinder
- Stopwatch or clock
- Water
- Stirrer
- Thermometer.

Here is what you should do:

1. Measure out 50 cm³ of water into each of three 100 cm³ beakers.
2. Use a thermometer to confirm that the water in the three beakers is at the same temperature. If it is not, allow the beakers to stand until the temperature is the same.
3. Label the three substances you have been given as A, B and C.
4. Weigh exactly 1 g of Substance A.
5. Pour Substance A into a beaker of water, start timing and stir until all of the substance has dissolved. Record the time it takes.
6. Repeat steps 4 and 5 for Substance B, and then for Substance C.
7. Record your results in the form of a table.

In this activity you carried out a fair test because all of the conditions under which each of the three substances dissolved were exactly the same.

Check your understanding

1. In the above activity:

   a) what was the variable?

   b) what conditions were kept constant?

> **Fun fact**
>
> The number of days in any given month is a constant but the number of days in each month varies over the year between 28 and 31.

Key terms

constant something that always has the same value

variable something that may take different values

fair test test in which all but one variable is kept constant

Writing a lab report

Experimental reports ⟩⟩⟩

Whenever scientists carry out their experiments, they record each section using a particular format. This is called a **lab report**. It is necessary for budding scientists to learn the correct method for writing reports.

Format for writing a report

Topic: Give the name of the experiment.

Date: State when you did it.

Aim: State what is being explored and make any prediction(s).

Apparatus and materials: Identify all the tools and items to be used. Include diagrams of how to set up the apparatus.

Method/procedure: Record the steps of the experiment:
- Use the past tense
- Write the steps in order
- Number the steps.

The method may include diagrams.

Observation/results: Record what happens – anything you observe – and any measurements. Use tables and graphs. You may include annotated diagrams.

Analysis and discussion: Discuss your observations. Explain what these show.

Precautions taken: Explain the steps you took to limit errors.

Conclusion: Draw inferences from the experiments.

FIG 1.8.1

Activity 1.8.1

Writing a report

Now that you know the format for writing a report on an experiment, see how well you can write up a report on your balloon experiment. Make sure that your report is neat and well laid out. You can use a computer if one is available.

It requires practice to become a scientist. Hence, in this course, there will be many opportunities for carrying out and **recording** experiments.

FIG 1.8.2

Check your understanding

1. Below are some pieces from lab reports. Say in which section they would come in a lab report.

   a) A container was filled with oil and another container was filled with water. A new iron nail was placed in each container and left for a week.

   b) A metre rule was used to measure the heights of all the plants.

   c) When the freezer was opened, it was found that all the containers were cracked.

   d) The water expanded when the ice was formed.

   e) Since both lids A and B were the same size, the heat had caused lid A to expand.

2. In a lab report, the Aim may include predictions. Which section tells whether a prediction is correct?

Key terms

lab report a report of a scientific investigation or experiment in a standard format

aim stating what is being explored and making any prediction(s)

apparatus the tools and equipment that scientists use

method a written record of the steps of an experiment

analysis an explanation or the reason for the method used and of the steps taken to limit errors

conclusion a deduction made based on information gathered

recording writing down information that can be used or seen again

Engineering design process

We are learning how to:

- use scientific skills in the engineering design process.

Engineering design process »»

The scientific method you learned about in Unit 1.6 is one way of approaching a problem but it is not the only way. The engineering design process (EDP) provides an alternative method which may be better suited to some scientific work.

There are several steps we can identify in each EDP:

- Defining the problem – what are you hoping to achieve?

- Carry out **research** – can you find out anything from resources such as books, the Internet or simply talking to people with particular expertise that will help you?

- Specific requirements – what equipment and materials are you going to need and where can you get them?

- Generating solutions – can you see different ways to solve the problem? Which way looks the best?

- Build a **prototype** – go with what you think is your best idea and build the apparatus you need.

- Test and redesign – try out your prototype. You might need to modify or redesign it before it works how you would like it to.

- Communicate results – providing accurate details of the work you carried out is an important part of science.

The following activity provides an opportunity for you to carry out the EDP.

FIG 1.9.1 Research is the first active step in the EDP

Activity 1.9.1

Using a fuel made from recycled material to replace charcoal

Traditionally coal stoves and jerk pans use charcoal to cook food. Charcoal is supplied in bags as different sized chunks. Here are two important facts which are the basis of your task:

- Although charcoal is made from wood, which is a renewable source of energy, its manufacture damages the environment and creates pollution.

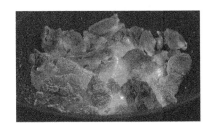

FIG 1.9.2 Charcoal provides the heat to cook food

- Every year as a society, we produce millions of tonnes of waste paper and cardboard. Much of this can be **recycled**.

1. **The problem**: You must manufacture an alternative fuel for coal stoves from waste paper.

2. **Research**: Look at the different brands of charcoal for sale in your local market or supermarket. Measure the sizes of the chunks. This will help you decide what size to make your briquettes. Look at the devices you can buy to compress wet paper into cubes or logs.

3. **Requirements**: Identify sources of waste paper and/or cardboard that can be used to make briquettes. Can you make, borrow or buy a device to compress wet paper?

4. **Solutions**: Can you think of some different ways of wetting and compressing your briquettes? What is the best way to dry them?

5. **Prototype**: Have you got something that you think is going to work? Choose what you consider your best briquette for testing.

6. **Testing**: How are you going to test your briquettes? It takes 4.184 J of heat energy to raise the temperature of 1 g of water by 1 °C. Perhaps you could heat a given mass of water with a given mass of briquette for a set time and measure the temperature increase?

   As a result of the test do you need to modify your production methods in any way?

7. **Report**: Write a report about what you did. You might record things like:
   - what you used to make your briquettes
   - how you made them
   - how your briquette performed compared to charcoal.

   You could take some photographs to illustrate your report.

You will have plenty of opportunities to apply the EDP as there is a science, technology, engineering and mathematics (STEM) activity at the end of each unit.

Check your understanding

1. At what point in the EDP would you:

   a) build a prototype?

   b) carry out research?

   c) modify the prototype?

   d) write a report?

Key terms

research find out using resources of different kinds

prototype working model or first attempt as a solution

recycle reuse an article or reuse the material it is made from

Scientific measurement and SI units

We are learning how to:
- state what measuring is and why we need to measure
- use standard units.

What is measuring and why do we measure?

Measuring is the means by which the size, mass, and/or temperature of some quantity of matter, or a period of time, is determined using **instruments** marked in **standard units**.

It is necessary to measure so that communication about any quantity of matter is clear.

The main **physical quantities** that can be measured are:

- Dimensions – the size of an object in terms of, for example, its length, width, height or depth

- Volume – how much space an object occupies

- Mass – how much matter an object contains

- Time – how long an event lasts

- Temperature – how hot an object is.

Units of measurement

We need to have standard units of measurement so that we can make useful comparisons. In science, we use mostly **SI units**. The different SI units are shown in Table 1.10.1.

It's not far to Montego Bay – it's about 250 pieces of string away.

FIG 1.10.1 How far is it to Montego Bay?

FIG 1.10.2 A thermometer

Physical quantity	SI unit	Symbol	Instrument/s
Length	Metre	m	Metre rule, tape measure, calliper
Mass	Kilogram	kg	Balance
Time	Second	s	Stopwatch, clock
Temperature	Kelvin	K	Thermometer

TABLE 1.10.1 Some SI units of measurement, their symbols and the instrument(s) used

Other units are also used, such as degrees Celsius (°C) for temperature and minutes for time.

Multiples and submultiples of SI units have names; for example:

1 kilometre = 1000 metres or 1 km = 1000 m

1 centimetre = 1/100 metre or 1 cm = 1/100 m

1 millimetre = 1/1000 metre or 1 mm = 1/1000 m

Centimetres are often used for the area and volume of everyday objects, as shown in Table 1.10.2.

The volume of a liquid is often measured in litres (l) and millilitres (ml). One millilitre (1 ml) is equal to one cubic centimetre (cm^3).

Physical quantity	Unit	Symbol
Area	Square centimetre	cm^2
Volume	Cubic centimetre	cm^3

TABLE 1.10.2

Activity 1.10.1

How good are you at estimating?

Here is what you need:

- Rulers
- Tape measures
- Metre rules
- Stopwatches
- Balances.

Here is what you should do:

1. Look around the classroom and select objects such as your textbook, your dictionary, your pencil case, your backpack and so on.

2. In your groups, test each other's skill at estimating the lengths and masses of the objects chosen.

3. Use an appropriate measuring instrument to check your estimates.

4. Back in class, discuss the techniques you used to estimate values and how good your estimates were.

Estimating is a useful skill, but our senses are not reliable. Each person's ability to judge a physical property is different. As a result, it is necessary to use specially designed measuring instruments and standard units that give us the information consistently and globally.

Check your understanding

1. Explain why standard units of measurement are needed.

2. How many millilitres are in one litre?

3. What instrument do we use in the laboratory to measure the volume of a liquid? What might we use in the kitchen at home?

4. What unit of mass is 1/1000 of a kilogram?

Fun fact

Yonge Street in Toronto, Canada, is the longest street in the world. It measures 1896 km (1178 miles).

Key terms

measuring using instruments to determine quantities

instrument a tool used for measuring

standard unit a value that can be communicated universally

physical quantity an amount that can be measured

SI units international system of units of measurement

estimate a calculated approximation

Scientific apparatus

We are learning how to:

- identify and draw some common apparatus used in school science lessons
- use a Bunsen burner.

Apparatus for scientists »

As with all workers, scientists have special tools for their jobs. These tools and equipment are called 'apparatus'. There are dozens of pieces of apparatus, each with its own unique use.

Activity 1.11.1

Getting familiar with apparatus

Here is what you should do:

1. Your teacher will show you some pieces of apparatus. See whether you can guess what the different pieces of apparatus are used for.

2. Practise drawing items of apparatus, and learn their names and uses.

Fig 1.11.1 shows some scientific apparatus.

Test tube rack

Safety glasses

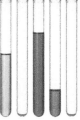
Reagent bottle

Bunsen burner

Evaporating dish

Flat-bottomed flask

Round-bottomed flask

Test tube

Filter funnel

Beaker

Electronic balance

Measuring cylinder

Stand and clamp

FIG 1.11.1 Some scientific apparatus

Each piece of apparatus shown in Fig 1.11.1 has a particular use and function:

- **Test tube or boiling tube** – for containing (may be for heating) small amounts of substances

- **Beaker** – for containing chemicals or collecting liquids
- **Measuring cylinder** – for measuring accurately the volume of liquids
- **Round-bottomed flask** – for preparation of gases if the process requires heating
- **Flat-bottomed flask** – for preparation of gases if no heating is required
- **Filter funnel** – for separating an insoluble solid from a liquid with the help of filter paper
- **Evaporating dish** – for evaporating a liquid from a solution
- **Electronic balance** – for weighing liquids or solids in a container
- **Reagent bottle** – for containing chemicals in liquid or powder form
- **Test tube rack** – for holding test tubes
- **Stand and clamp** – for supporting apparatus during experiments
- **Bunsen burner** – to provide a flame for heating
- **Safety glasses** – to protect your eyes while experimenting.

Fun fact

Mechanical lighters were used before matches were invented.

Key terms

Bunsen burner
apparatus which provides heat and named after its inventor

luminous flame
a warm, yellow flame burning with incomplete combustion

non-luminous flame
a bright, very hot, blue Bunsen flame which burns with complete combustion, producing water and carbon dioxide

Bunsen burner

Your teacher will demonstrate and explain the correct procedure used to light a **Bunsen burner**, and will show and explain the difference between a **luminous flame** and a **non-luminous flame**.

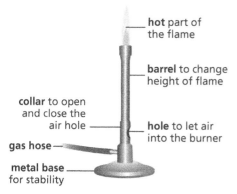

hot part of the flame

barrel to change height of flame

collar to open and close the air hole

hole to let air into the burner

gas hose

metal base for stability

FIG 1.11.2 Parts of the Bunsen burner and their functions

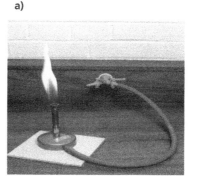

a)

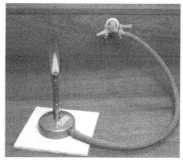

b)

FIG 1.11.3 Bunsen burner flames: **a)** luminous and **b)** non-luminous

Activity 1.11.2

Understanding apparatus

1. Find out who the Bunsen burner was named after.

2. Explain what you would use a Bunsen burner for. When might it be dangerous to use a Bunsen burner?

Presenting data

Data is another word for information. Scientists often collect data of different kinds and display it in appropriate ways. Collecting data often involves taking measurements like length, mass, volume and temperature.

Here is an example of how a scientist gathered information about dog whelks. These are animals that live on rocky sea shores.

The scientist measured the shell length of 25 dog whelks. Here are the measurements correct to the nearest millimetre:

27, 22, 21, 25, 24, 22, 26, 25, 23, 26, 25, 23, 28, 24, 26, 27, 26, 27, 26, 24, 25, 28, 24, 26, 25

This is called **raw data**. She organised the data into a **tally chart** (Table 1.12.1). This allows her to see the number or **frequency** of animals at each different shell length.

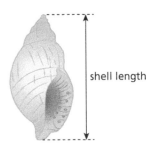

FIG 1.12.1 A dog whelk

shell length

Shell length / mm	Tally	Frequency
21	I	1
22	II	2
23	II	2
24	IIII	4
25	IIII	5
26	IIII I	6
27	III	3
28	II	2
	Total	25

TABLE 1.12.1

The pattern of shell length can be seen more easily if the data in the frequency table is presented in some way. Ways of displaying data include a pictograph, **bar graph**, histogram, pie chart and **line graph**.

This type of data is suitable for display as a bar graph.

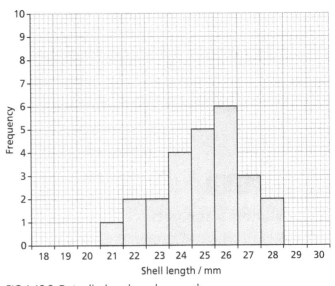

FIG 1.12.2 Data displayed as a bar graph

Activity 1.12.1

Collecting data

Here is what you need:

- 30 cm ruler.

Here is what you should do:

1. Go into the school field or an undeveloped area near to where you live.

2. Find a population of living things that you can measure. For example, snails or leaves, flower stems, flowers, etc.

3. For a sample of 25 of your chosen living thing, measure the same item for each of them.

4. Record your raw data.

5. Use your raw data to compile a frequency table.

6. Present the data in your frequency table as a bar graph.

Key terms

raw data data before it is organised in any way

tally chart means of adding the numbers of items at different values

frequency the number of items of a particular value

bar graph method of displaying data as a series of vertical or horizontal bars of different heights

line graph method of displaying data as a line joining different values

Check your understanding

1 Table 1.12.2 shows the loss in mass, due to the evolution of carbon dioxide gas, during a chemical reaction between calcium carbonate and hydrochloric acid.

a) Copy and complete Table 1.12.2 by calculating the total mass lost at each time interval.

b) Display this data as a line graph.

Time / minutes	Mass of reaction vessel and contents / g	Total mass lost / g
0.0	190.45	0.00
0.5	189.89	
1.0	189.49	
1.5	189.22	
2.0	189.01	
2.5	188.87	
3.0	188.73	
3.5	188.63	
4.0	188.53	
4.5	188.43	
5.0	188.36	
5.5	188.31	
6.0	188.27	
6.5	188.26	
7.0	188.25	
7.5	188.25	
8.0	188.25	

TABLE 1.12.2

Making scientific drawings

We are learning how to:

• draw scientific diagrams.

Scientific diagrams ⟩⟩⟩

When scientists record the apparatus used and observations made in an experiment, they often need to draw diagrams. Producing diagrams in science does not require special art skills. There are, however, some basic guidelines on scientific drawings, which are outlined below.

• All scientific drawings must be as simple as possible but also as true to life as possible.

• Drawings must be done in pencil – no coloured pens or markers should be used.

• Sharp pencils and a ruler, eraser and sharpener are needed.

• Various pencil techniques are used for natural objects. You should draw:

   o smooth-edged specimens using smooth unbroken lines

   o specimens of wool, cotton or hair with fuzzy edges

   o underlying structures with broken lines (to indicate their shape).

• No shading is done. Other techniques are used:

   o Stippling (dotting)

   o Streaking (lines in one direction)

   o Cross-hatching (lines crossing each other).

FIG 1.13.1 This drawing of a plant uses colour and shading; these are not needed for scientific diagrams

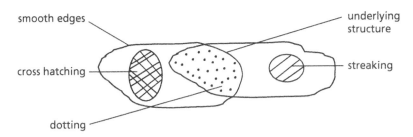

FIG 1.13.2 Some of the drawing techniques used in science

• **Symbols** may be used if their meaning is made clear.

Activity 1.13.1

Drawing a leaf

Here is what you need:

- A simple-shaped leaf that is neither too small nor too large

- Drawing materials.

Here is what you should do:

1. Place the leaf on the desk so that you do not have to move it while you draw.

2. Draw the outline of your leaf in the left side of your page. Your drawing should have no sketchy edges and look as smooth and sharp as in Fig 1.13.3.

FIG 1.13.3 Outline of a simple leaf

3. Draw the labels to the right of your diagram. Put the title of the diagram, the view and the magnification underneath the diagram. Use capitals for the title. Do not discard the leaf.

Your teacher will assess your drawing and guide you to complete the diagram, which may look like the leaf in Fig 1.13.4.

FIG 1.13.4 Drawing of a simple leaf

Check your understanding

1. A student drew this picture of a flower. List three good points about this drawing and then list three bad points.

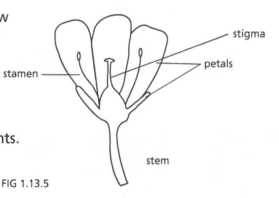

FIG 1.13.5

2. Explain why label lines should be straight.

> **Fun fact**
>
> The falling of autumn leaves is extremely important to the ecological balance of the environment. The dropped leaves help to form the organic layer of the soil surface, and the nutrients in the leaves are recycled by decomposers.

Key term

symbol a character or drawing used to represent a situation

Labelling scientific diagrams

We are learning how to:
- label scientific diagrams
- work out the magnification of a diagram.

Annotating a diagram 》》

Scientific drawings need to be **annotated** – labelled with descriptive words. There are some rules to follow when labelling a diagram.

- **Title**:
  - ○ indicates what the diagram represents
  - ○ is best written in CAPITAL letters
  - ○ is best placed at the bottom centre of the drawing.
- **Labels**:
  - ○ identify the various parts of the drawing
  - ○ are written in pencil
  - ○ are either all small letters or all CAPITAL letters, but not a mixture.
- **Labelling lines**:
  - ○ are drawn horizontally wherever possible
  - ○ are drawn in pencil using a ruler
  - ○ should connect to the drawings
  - ○ do not have arrowheads or dots at either end
  - ○ end on the part or in the space they identify
  - ○ do not cross over each other.
- For biological drawings of natural objects, the **magnification** and the view (for example, plan view – from the front, side view – in profile, or cross-section – cut through and showing the cut edge) of the drawing are written close to the title.

Activity 1.14.1

Labelling your leaf diagram

Here is what you need:
- Your leaf diagram from Activity 1.13.1.

Here is what you should do:

- Label your leaf diagram. (Your teacher will guide you to do this correctly.)

When you have labelled your leaf diagram, it should look like the following:

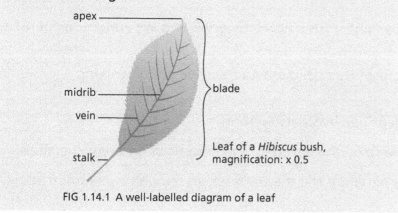

apex

midrib

vein

blade

stalk

Leaf of a *Hibiscus* bush, magnification: x 0.5

FIG 1.14.1 A well-labelled diagram of a leaf

Calculating the magnification

If the diagram is bigger than the **specimen**, do you think the magnification will be larger or smaller than 1? (Hint: the actual specimen is always rated 1.)

Place the specimen over the drawing and judge how many times the specimen can fit.

To find out the magnification of your diagram:

1. Measure the length or the width of both the specimen and your drawing.

2. Divide the dimension of your drawing by that of the specimen. The result is the magnification.

$$\text{Magnification} = \frac{\text{length of drawing}}{\text{length of object}}$$

> **Fun fact**
>
> It is not necessary for frogs to drink water, because frogs absorb water through their skin.

Key terms

annotated having important or explanatory notes

magnification the scalar representation of an actual specimen

specimen the item or material under examination

Check your understanding

1. Fig 1.14.2a represents a specimen (a dissected frog). Fig 1.14.2b is a diagram of the specimen. Work out the magnification.

a)

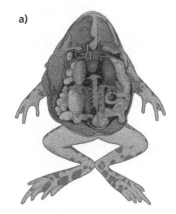

b)

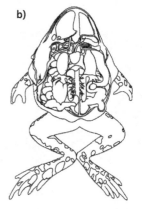

FIG 1.14.2 **a)** A dissected frog

FIG 1.14.2 **b)** A diagram of the dissected frog

Review of Science and scientific processes

- The laboratory is a potentially dangerous place if materials and apparatus are not used correctly.

- Laboratory rules are there to protect individuals and provide a safe working environment.

- Dangerous substances are identified by safety symbols.

- Safety symbols provide guidance on the nature of the hazards of different chemicals.

- Science is the systematic study of nature and the environment through a scientific method.

- There are many different branches of science.

- Technology is the appliance of science in practical ways to improve life.

- There are many famous scientists in different branches of science.

- Some scientists have been honoured by having laws, units or elements named after them.

- Jamaica has produced a number of famous scientists over the years.

- The scientific method involves making careful observations, recording and analysing data.

- Scientific skills are essential for accurate results.

- A lab report should be written in a particular format indicating the apparatus and materials used, what was done, what was observed or measured and what conclusions can be made.

- To provide meaningful results an experiment must involve fair testing.

- The engineering design process (EDP) provides an alternative way to carry out scientific work.

- EDP involves identifying the aim, carrying out research, sourcing equipment and materials, building and testing a prototype, making modifications and writing a report.

- Measuring is determining some physical quantity of matter by using instruments marked in standard units.

- Measuring and the use of standard units makes it possible to communicate measurements clearly.

- There is an internationally agreed set of units called 'S.I. units' which include the metre (m), kilogram (kg), second (s) and kelvin (K).

- Scientists use apparatus to carry out experiments.

- Apparatus consists of containers, measuring devices and other objects made of glass, metal and plastic.

- Data is another word for information.

- Raw data is data as it is collected and before it is organised in any way.

- Tally charts and frequency tables are used to organise data.

- Data can be displayed in various ways including a bar graph and a line graph.

- Drawing scientific specimens requires a sharp pencil.

- Different techniques are used to draw natural objects including unbroken and broken lines, stippling, streaking and cross-hatching.

- Scientific drawings must be labelled and given a meaningful title.

Review questions on Science and scientific processes

1. State the meaning of each of the following warning signs.

a) b) c) 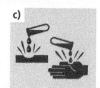 d) e)

FIG 1.RQ.1

2. Write a sentence to explain each of the following laboratory rules.

a) Wear eye protection when carrying out experiments.

b) Don't eat or drink in the laboratory.

c) Don't leave coats or bags in the middle of the floor.

d) Report all accidents no matter how trivial they seem to you.

3. Write one sentence to describe each of the following branches of science.

a) Astronomy

b) Zoology

c) Meteorology

d) Metallurgy

4. a) Explain the difference between science and technology.

b) Give one example of how technology has improved your life.

5. a) Which famous scientist gave their name to the unit of:

    i) force?   **ii)** electric current?   **iii)** power?

  b) What famous scientist(s) are the following elements named after?

    i) Curium   **ii)** Einsteinium

6. Here are some different parts of the engineering design process:

  identifying the problem   research   obtaining materials and equipment

  making a prototype   testing and modifying   reporting

  In which of these parts should you:

  a) make a list of the apparatus and materials you are going to use?

  b) collect data?

  c) say what you did?

  d) make a model or plan a process?

7. What is the S.I. unit used to measure:

  a) length?   **b)** time?   **c)** mass?

8. Briefly explain the significance of each of the following stages in planning an experiment.

  a) Hypothesising

  b) Designing the experiment

  c) Collecting data

  d) Analysing data

  e) Drawing conclusions

9. Here are pictures of some pieces of apparatus.

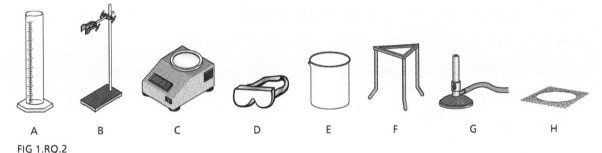

  A   B   C   D   E   F   G   H

FIG 1.RQ.2

Which of these pieces of apparatus:

a) provides protection from injury when carrying out experiments?

b) measures mass?

c) is a source of heat?

d) measures volumes of liquids?

10. a) Name five quantities that a scientist might measure.

   b) For each, state the units that would be used to express each quantity.

   c) For each, state the instrument used to make the measurement.

11. The following shows the heights of 24 plants. The plants were measured to the nearest 2 cm.

   36, 42, 40, 44, 38, 38, 40, 40, 34, 40, 42, 40, 38, 36, 34, 36, 40, 38, 42, 44, 38, 36, 42, 40

   a) Construct a table to show the tally and frequency for each height up to 48 cm.

   b) Represent this data as a bar chart.

What are STEAM activities?

STEAM stands for **S**cience, **T**echnology, **E**ngineering, **A**rt and Mathematics. The purpose of STEAM activities is to demonstrate how the science that you learn in the classroom can be applied to solve problems in many different fields.

At the end of each unit of this series you will find one or more STEAM activities designed to provide opportunities for students to apply the different subject knowledge. For each activity you will work in a group of 3 or 4 students and are expected to make a significant contribution towards the team effort.

FIG 1.SIP.1 STEAM activities provide opportunities for working in teams

There are several stages to every STEAM activity. The exact details will be different for each specific task, but we can summarise the stages in general terms.

1. Defining the problem – This tells you exactly what is required of the group. It will often be a problem that needs solving. It may involve making something or carrying out tests of some kind and observing results.

2. Research – In the remaining STEAM activities at the ends of the units in this book you will be expected to review the subject information given in the unit and supplement this by looking at other sources such as reference books and the Internet.

3. Requirements – You need to think about what you will need and make lists. You may need a list for materials and another list for tools and equipment. You might also need to obtain advice from someone outside your group, such as your teacher.

4. Agree on a solution – You are working as one student in a team of 3 or 4. The members of the team will probably suggest more than one solution to the problem. If so, then team members need to analyse and evaluate the suggestions of the different members and come to an agreed solution.

5. Building a prototype – This provides the team members with an opportunity to show their manual skills. The task should be organised in such a way that each member of the team makes a contribution. It may be useful if one person takes on the role of team coordinator.

6. Testing – Once the prototype is built, it needs to be tested or evaluated. It is often useful to have people from outside the team, who have not been closely involved in the planning and building, to review your work. A fresh pair of eyes can sometimes pick up things missed by people who are closely involved. Consider the comments made by your reviewers and modify what you have made in the light of them if you think it is appropriate.

7. Reporting back – The final task is to compile and deliver a report to explain what you did and to show your end product. This can take various forms, such as a PowerPoint presentation or a demonstration of what you have designed and built. If possible, take pictures at different stages of construction and testing on your cell phone. Illustrated reports are always more interesting than just writing and description.

1. Your teacher has asked you to devise a game to demonstrate the concept of random motion.

2. Random motion is the movement of an object in directions which cannot be predicted and show no pattern, i.e. they are random.

3. To demonstrate random motion, you will need:

   - Something that generates random data.

FIG 1.SIP.2 Movement in six directions

An object can move in six directions. A dice has six numbers. Can you see how a dice might be used to generate random data?

   - Something that can be moved in a random way.

Fig 1.SIP.3 shows the board used for three-dimensional tic-tac-toe. Does this give you any ideas how an object may be show to move in random directions?

FIG 1.SIP.3 3D Tic-tac-toe

4. The members of your team need to discuss their ideas and decide on a solution.

5. To build your prototype, make a list of material and tools you are going to need. Make full use of what may be available. For example:

   - You might decide to you want a dice showing six directions. Instead of starting from scratch, paint out the numbers on an existing dice and paint on letters representing directions: F, B, R, L, U and D.

   - The levels on a three-dimensional board don't have to be clear plastic. They can be made of cardboard mounted on four pieces of dowel. Your object can be a marble or a pebble.

Be sure to take photographs as you build your prototype. These will be useful when you compile your report.

6. Once your model is complete, ask some friends from outside the group to test it. You will need to explain how it works. For example:

   - put the object in the middle of the middle level
   - roll the dice
   - move the object one place in that direction.

7. Your final task is to write a report describing what you have done. This could be a PowerPoint presentation. Use photographs to illustrate your report.

Unit 2: Nature of matter

We are learning how to:
- investigate some properties of matter
- show that matter occupies space
- show that matter has mass

What is matter? »

Matter is everything around you that has mass and occupies **space**. Matter can be in one of three forms: a **solid**, a **liquid** or a **gas**. For example, water and gasoline are liquids at room temperature, and rocks, iron and glass are solids. They all occupy space and have mass. Air is a gas, but even air has mass and occupies space, even though its mass is very small.

a) The sea

b) Gasoline being pumped into a car

c) Rocks

d) Iron bars

e) Glass used in a building

FIG 2.1.1 Matter is all around us, in many different forms

Differences between solids, liquids and gases

Solids are generally rigid, hard and heavy. A concrete block is a solid: it is **rigid**, which means you cannot change its

> **Key terms**
>
> **matter** everything possessing mass and volume
>
> **space** the volume occupied by matter
>
> **solid** the form of a substance with a rigid structure and set volume
>
> **liquid** the form of a substance with no rigid structure but a set volume

shape easily, it is hard and it is heavy. However, not all solids are rigid, hard and heavy (see Fig 2.1.2).

Liquids are not rigid. They can **flow** or be poured. Some solids can be poured, but they are not liquids (see Fig 2.1.3a). Liquids are not usually shiny like solid metals, but one metal is a liquid at room temperature (see Fig 2.1.3b).

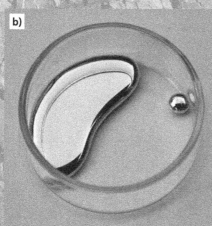

FIG 2.1.3 **a)** Granulated sugar can be poured, but it is not a liquid **b)** Mercury is a metal and is shiny, but it is a liquid

Gases are very light and they are not rigid. Many gases are invisible.

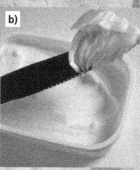

FIG 2.1.2 **a)** Polystyrene is not heavy, but it is solid **b)** Margarine is not rigid, but it is solid

FIG 2.1.4 **a)** Air is invisible, so how do we know it is there? **b)** We can see these gases used on stage during a rock show, and we can see that they flow.

Check your understanding

1. List two properties for each of:

   a) solids

   b) liquids

   c) gases.

2. We know water can exist in all three states of matter. Name the three states of matter of water.

Key terms

gas the form of a substance with neither a rigid structure nor a set volume

rigid firm and set in place

flow continuous movement of a fluid material

Properties of gases

We are learning how to:

- show that gases are matter
- show that gases have mass and occupy space.

Gases ⟩⟩⟩

Gases are all around us, in the air. We cannot see most gases, but we can investigate their properties.

Activity 2.2.1

Investigating the mass of a gas

Here is what you need:

- Metre rule
- Drawing pin
- String
- Two identical balloons.

Here is what you should do:

1. Tie three pieces of string to the metre rule: one piece hanging down from each end of the rule, and one placed exactly in the middle of the rule so that you can hold the string and balance the rule (see Fig 2.2.1).

2. Blow both balloons up to a good size, but do not overinflate them. They should both be blown up to the same size and a knot tied at the neck so they do not deflate.

3. Tie one balloon to the string at one end of the rule, and the other balloon to the string at the other end of the rule.

4. Hold up the string attached to the middle of the rule, so that the rule is balanced.

5. Pierce one balloon with the pin (keep hold of the pin).

6. Is the rule still balanced? What observation did you make? What conclusion can you draw?

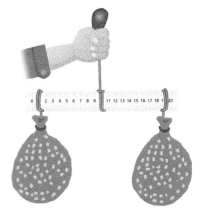

FIG 2.2.1

In Activity 2.2.1, an equal amount of air was blown into each balloon. This explains why the metre rule was balanced. When one balloon was pierced, the air in that balloon was released. Immediately, the rule went off balance as the inflated balloon went lower. This indicated that the inflated balloon had more mass than the one with no air.

Since the inflated balloon had more mass than the pierced one, we can conclude that air must have mass.

The experiment can be repeated for gases other than air, with similar results. All gases have mass.

Activity 2.2.2

Investigating the volume of a gas

Here is what you need:

- Trough or large flat-bottomed bowl
- Water
- Waterproof marker pen
- Balloon.

Here is what you should do:

1. Pour water into the trough up to about half-full and mark the level on the side of the trough.

2. Inflate the balloon and tie it at the neck.

3. Place the balloon gently into the trough. Does it sink?

4. Push down on the balloon. Can you make it go deeper into the water?

5. What do you observe about the water level as you push down on the balloon?

6. Can you explain what is happening?

7. Do you think that the water would behave the same way if the balloon were filled with any other gas?

In Activity 2.2.2, the inflated balloon had to be pushed down into the water. The water level rose up the side of the trough as you pushed. The deeper you pushed the balloon, the higher the water level rose. This means that a greater amount of space was needed to contain both the balloon of air and the water. This means that the air inside the balloon takes up some space. We say that the air has a **volume**.

The experiment can be repeated for any gas, not just air. This means that all gases have volume.

Check your understanding

1. Describe how you could show that the gas oxygen has mass and volume.

2. Can you think of anything that does not have mass or volume?

Fun fact

All the air above us in the atmosphere has mass and volume, which means that the air is pushing down on us constantly. This is what weather forecasters mean when they talk about 'atmospheric pressure'. We are used to this pressure, so we usually do not feel it.

Key term

volume space occupied by a three-dimensional object or substance

Density

We are learning how to:

- compare the masses of similar amounts of different materials.

Density »»»

It doesn't make sense to say that aluminium is lighter than lead because a big piece of aluminium will be heavier than a small piece of lead. To make a meaningful comparison we need to have pieces of aluminium and lead which are equal in size.

Density allows us to compare equal volumes of different materials.

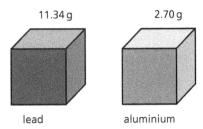

11.34 g   2.70 g

lead   aluminium

FIG 2.3.1 A cube of lead of sides 1 cm has more than four times the mass of a similar-sized cube of aluminium

Activity 2.3.1

Comparing the densities of different materials

Work in groups for this activity.

Here is what you need (for each group):

- Equal-sized blocks of different materials such as wood, sponge, glass, cork, stone, metal, etc.

- Balance.

Here is what you should do:

1. Hold each block in the palm of your hand in turn and say whether you think it feels light or heavy, or somewhere between.

2. Arrange the blocks according to how heavy they feel starting with the one you think is the lightest.

3. Write down the names of the materials in the order you have put them.

4. Weigh each of the materials on a balance. Write down the weights in the form of a table like the one below.

Material	Mass in g	Order according to mass

5. When you have weighed all the materials, write the order of their masses starting with the material that had the lowest mass.

6. Compare the order in which you wrote the names of the materials according to how heavy they felt with the order of their masses.

7. How good were you at sorting the materials according to how heavy they felt?

The mass of a substance is the amount of matter which it contains. When we compare the masses of blocks of the same size, but of different materials, we are comparing the mass per volume, which is called **density**.

Gases have very low densities because the particles in a gas are spread out. Even a large volume of gas has a very small mass.

Liquids have higher densities than gases because the particles in a liquid are much closer together. The metal mercury is a liquid at room temperature and has the highest density of any liquid.

Solids have the highest densities of all because the particles in a solid are closely packed together in a lattice.

Materials that have a density less than water, like cork, will float while materials that have a density more than water, like lead, will sink.

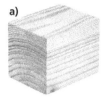
a)

b)

c)

d)

e)

FIG 2.3.2 Blocks of different materials: a) wood, b) glass, c) sponge, d) stone, e) metal

Check your understanding

1. Here is what happened when equal-sized blocks of materials were placed in a container of water.

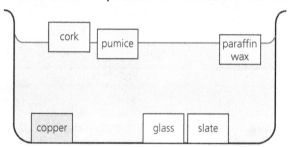

FIG 2.3.3

a) Which of these materials would feel light if you held them in your hand?

b) Which of these materials would feel heavy if you held them in your hand?

c) Aluminium is more dense than slate but less dense than glass. Predict whether a block of aluminium would float or sink in water.

Fun fact

The densest known element is a metal called osmium. A cube of osmium of sides 1 cm has a mass of 22.59 g. This is twice the mass of a similar cube of lead.

Key term

density the mass of a particular volume of a material

States of matter

We are learning how to:

- identify and differentiate between states of matter.

Matter and non-matter ⟩⟩⟩

Matter – or stuff – is all around us. But what exactly is matter, how does it exist, and does non-matter also exist?

Activity 2.4.1

Classifying forms of matter

We can classify (group) things in many ways.

Here is what you should do:

1. Look at the pictures in Fig 2.4.1. Choose some ways of classifying the items shown. Try to be scientific.

2. What do all the things have in common?

3. What are the different features?

4. Discuss with each other the reasons for your classification.

FIG 2.4.1 Examples of forms of matter

States of matter

All matter is made up of tiny units called **particles**. All matter has mass and volume. But matter exists in different forms. One way of classifying matter is by the state in which it exists. These states are: **solid**, **liquid** and **gas**.

We can determine **states of matter** by observing physical properties. Physical properties are features or characteristics that can we can observe without changing the substance.

- **Solids** have a definite volume and a **rigid** structure, which makes a definite shape.

- **Liquids** have a definite volume but no rigid structure, so they keep no definite shape. They take the shape of the part of a container that they are in.

- **Gases** have neither definite volume nor shape.

Liquids and gases are both **fluids** – they can flow – because their particles are able to easily slide over each other.

The properties of solids, liquids and gases and the arrangements of the particles in them are shown in Fig 2.4.2.

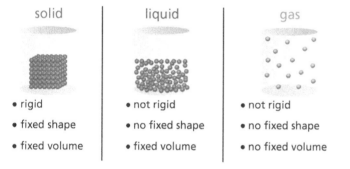

solid	liquid	gas
• rigid	• not rigid	• not rigid
• fixed shape	• no fixed shape	• no fixed shape
• fixed volume	• fixed volume	• no fixed volume

FIG 2.4.2 Arrangement of particles in a solid, liquid and gas, and the main properties of each state

Check your understanding

1. What properties do the following have in common:

   a) Solids and liquids?

   b) Liquids and gases?

2. Explain the difference between 'rigid' and 'fluid' and give examples of objects with these properties.

3. Write a description of the difference between a gas and a liquid, for someone who knows no science.

> **Fun fact**
>
> Some things exist that take up neither space nor volume; for example, you cannot measure 5 kilograms of shadows or 2 litres of light.

Key terms

particle tiny bit or quantity of matter

solid the form of a substance with a rigid structure and set volume

liquid the form of a substance with no rigid structure but a set volume

gas the form of a substance with neither a rigid structure nor a set volume

state of matter the form in which matter exists

rigid firm and set in place

fluid form of a substance that can be continually deformed (liquid or gas)

Particles of matter

We are learning how to:

- explain that bonds between particles determine the shape and volume of each state of matter.

Particles 〉〉

What determines the shape and volume of objects in different states of matter?

Activity 2.5.1

Investigating objects

Here is what you need:

- Stones
- Beaker of water
- Inflated balloon
- Empty bottle
- Piece of wood
- Sponge
- Baby powder.

FIG 2.5.1 Materials to investigate

Here is what you should do:

1. Examine the objects. Which objects:

   a) have a fixed shape?
   c) can be poured?
   b) are easily broken or separated?
   d) can be picked up and held easily?

2. What about what is inside the balloon and the bottle?

3. List some physical properties of each object.

Bonds between particles

The tiny particles that make up matter are held together by forces called **bonds**. The way in which the particles are arranged determines whether a substance is solid, liquid or gas.

Solids can be hard like a rock, soft like fur, big like an asteroid, or powdery. Strong bonds hold the particles of solids together and make the solids tightly packed (see Fig 2.5.2a). Baby powder is made from ground-down rock (talcum powder); tiny pieces of solid can be seen in the powder, under a microscope. Solids maintain their shape and volume.

Liquids do not keep any set shape. The bonds holding the particles of a liquid together are weaker, allowing the particles to move (see Fig 2.5.2b). You cannot cut the liquid in a beaker into two pieces, because the liquid flows back to fill the space.

Gases cannot keep any shape and cannot stay put. Gas particles move continually, because there are almost no bonds between them (see Fig 2.5.2c). As a result, gases can take the shape of any large container and can also be **compressed** (squashed) into a tiny space.

Particles in a solid are held in position because the particles do not have enough energy to overcome the forces of attraction. Particles in a liquid move and so have more kinetic energy, but they still remain in touch with each other. Particles in a gas have much more kinetic energy, so they have sufficient energy to overcome the forces of attraction and move away from each other.

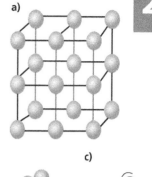
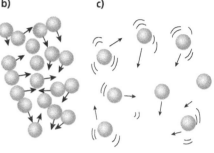

FIG 2.5.2 Bonds in states of matter: a) solid, b) liquid, c) gas

Fun fact

Of all solids, diamond has the strongest bonds between its particles. This makes it the hardest, strongest natural material. Drills that cut through rock have tips containing diamond.

Check your understanding

1. Write a sentence about each of the objects you investigated in Activity 2.5.1. In each sentence, try to explain how the bonds between the particles of the object cause it to have an observed property. For example:

   'The stones are hard and difficult to break, because the bonds between the particles are very strong.'

   Remember that the balloon and the empty bottle are containers of gas. Think carefully about the sponge.

Key terms

bond forces of attraction between particles, holding matter together

compress squeeze into a small space

Properties of solids, liquids and gases

We are learning how to:

- explain how properties of solids, liquids and gases are related to the way in which their particles are arranged and move.

Properties 〉〉〉

All materials are composed of particles. The different properties of solids, liquids and gases are the result of the way in which these particles are arranged and how they move.

The particles in a solid are held in fixed positions. They can vibrate but cannot move, so solids have a **fixed shape**. A solid block is the same shape whatever container it is placed in.

The particles in a liquid and in a gas can move about. A liquid and a gas will take the shape of the container in which they are placed.

Solids cannot be **poured** from one container into another but liquids and gases can. The fixed shape of solids means that they cannot form a stream in the same way as liquids and gases.

FIG 2.6.1 A solid block has a fixed shape

FIG 2.6.2 Liquids can be poured

Is it possible to compress or squash solids, liquids and gases?

Here is what you need:

- Plastic syringe × 3

- Short length of wooden dowel that fits inside one of the syringes

- Water.

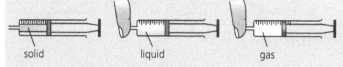

| solid | liquid | gas |

FIG 2.6.3 Compressing solids, liquids and gases

Here is what you should do:

1. Place the wooden dowel in the syringe and try to force the piston down the barrel.

2. Draw water into the syringe until it is nearly full. Place a finger over the end of the syringe and try to force the piston down the barrel.

3. Draw air into the syringe until it is nearly full. Place a finger over the end of the syringe and try to force the piston down the barrel.

4. Which of the three states, solid, liquid and gas, was easy to squash and which was difficult or impossible?

FIG 2.6.3

The particles in solids and liquids are close together so they cannot be **squashed** much closer together. The particles in a gas are far apart so it is possible to squash them closer together.

> **Fun fact**
>
> Sand is composed of many small particles of solid materials. We can pour sand from one container into another. Is this the same as pouring a liquid or a gas?

1. Create a table to compare solids, liquids and gases using the following characteristics: volume, shape, arrangement of particles, movement of particles, ability to be squashed. Include examples under the appropriate headings.

Key terms

fixed shape a shape that doesn't change

pour flow in a steady stream

squash make smaller

The effect of heat on matter

We are learning how to:

- explain what happens to a substance as it changes from solid to liquid and back to solid
- explain the link between the temperature and the state of matter.

Matter can melt »

We can investigate what happens when substances **melt**.

Activity 2.7.1

Room temperature – cool or warm?

Here is what you need:

- Shallow dishes such as half of a Petri dish × 4
- Ice cube
- Butter
- Lipstick
- Candle.

Here is what you should do:

1. Look at the four materials. In what state are they?

2. Place each material in a shallow dish.

3. Leave the dishes on the bench for five minutes.

4. Look what has happened to each material and write down what you observe.

Activity 2.7.2

Investigating melting

Here is what you need:

- Beaker of ice
- Butter
- Lipstick
- Candle wax
- Stopwatch
- Tin lids × 3
- Tongs
- Thermometer
- Cobalt chloride paper
- Water bath.

Here is what you should do:

1. Place the thermometer in the beaker of ice and read its temperature.

2. Place the beaker of water above the water bath and heat it for 5 minutes. In what state is the ice now? What is its temperature?

FIG 2.7.1 Experiment with melting ice

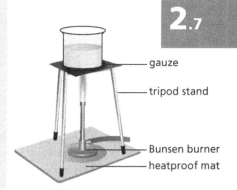

3. Touch the ice, or what it has changed into, with the cobalt chloride paper. What colour change takes place?

4. Place a small amount of butter on a tin lid and hold it over the water bath using tongs. Using a stopwatch time how long it takes before the butter starts to melt.

5. Repeat step 4 for lipstick and for candle wax.

6. Which of the four materials melted quickest?

7. Which of the four materials melted slowest?

FIG 2.7.2 Apparatus set up for heating tin lids

gauze
tripod stand
Bunsen burner
heatproof mat

All four materials in the Activities started as solids. At room **temperature** only the ice melted. The pink colour of the cobalt chloride paper indicated that it was water. With a slightly warmer temperature over the water bath, the butter melted, while it took more **heat energy** to melt the candle and the wax. This shows that each substance has a different melting point. When the temperature of a substance reaches its melting point, the particles of the substance have gained enough energy to move faster and further apart. As the particles move apart, they overcome the bonds between them and become fluid.

a)

b)

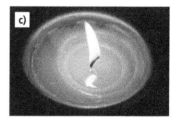

c)

Each substance also has a different solidifying point. Although room temperature is high enough to melt ice, it is sufficiently cool to allow candle and wax to **solidify**. Solidifying indicates that the particles of a substance have lost enough energy to slow them down and move closer to each other. As the particles move closer, the bonds overcome the movement energy of the particles and they form a rigid structure – they become solid.

FIG 2.7.3 Substances melting:
a) butter, b) lipstick, c) candle wax

Check your understanding

1. Explain why butter can go runny in a warm room. Use the words 'melt', 'temperature', 'heat energy' and 'particles' in your explanation.

2. Explain what happens if you then put the butter in the fridge. Use the words 'temperature', 'energy', 'solidify' and 'particles' in your explanation.

Key terms

melt when a substance changes from a solid to a liquid

temperature a measure of how hot an object is

heat energy energy produced based on a change in temperature

solidify when a liquid changes form to a solid

Melting and solidifying

We are learning how to:

- explain that changing a substance from solid to liquid or from liquid to solid is a reversible change
- classify these changes of state as physical changes.

Reversible changes 》》

Activity 2.8.1

Investigating ice

When a substance changes from solid to liquid, does it remain the same substance?

Here is what you need:

- Piece of cobalt chloride paper
- Zip-lock bag
- Water.

Here is what you should do:

1. Fill the zip-lock bag with water.
2. Measure the temperature of the water.
3. Place the filled bag into the freezer of your refrigerator.
4. Collect zip-lock bag of water from the freezer. Is the water still in the liquid state?
5. Measure the temperature and compare it with the temperature before you placed the bag in the freezer. Has it increased or decreased? Why has it?
6. Allow the ice to melt a little and then dip a piece of cobalt chloride paper into the liquid. Does it turn pink? What does that show?
7. Has melting and freezing changed the substance?

FIG 2.8.1 The ice in this berg is gradually melting to water

In a freezer, water exists as ice because the temperature is lower than 0 °C. When ice is exposed to room temperature, it melts because room temperature is higher than the temperature of the freezer.

At temperatures between 0 °C and 100 °C, water is in the liquid state. At any point in that range, if water is put in the freezer, the low temperature in the freezer causes the water to lose energy. Heat energy flows from the water

FIG 2.8.2 Coconut oil also solidifies at low temperatures

to the inside of the freezer causing the water to **solidify**. ('**Freezing**' is a type of 'solidification'.)

Melting and solidification are both **changes of state**. These are **reversible** processes. The change can go either way, depending on the temperature and the flow of heat energy.

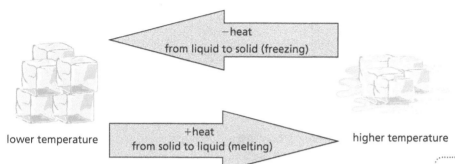

lower temperature   higher temperature

FIG 2.8.3 The effect of change of temperature on melting and freezing

Neither freezing nor melting changes the composition of the substances. Water remains water. Melting and solidification are **physical changes**, because the substance remains chemically unchanged even though it goes through a change of state.

Check your understanding

1. Write down these pairs of things and draw an arrow between them to show which way the heat energy will flow. (Hint: think about what temperature they are.)

Cold drink	Hot sunny surroundings
Hot cup of tea	Normal temperature room
Pie at room temperature	Fridge
Ice cream from freezer	Normal temperature room

TABLE 2.8.1

2. In which case in question 1 will there be a change of state?

3. Explain what we mean by a reversible change.

FIG 2.8.4

Fun fact

The Fahrenheit scale was created in the early 1800s. The zero point measures the temperature of a mixture of ice and salt. The freezing point of water on this scale is 32 while the boiling point is 212.

Key terms

solidify when a liquid changes to a solid

freezing when a substance changes from a liquid to solid because of a change in temperature

change of state the transformation of the form of a substance

reversible the ability for a substance to return to its original form when it is transformed

physical change a transformation that does not alter the chemical properties of a substance

Boiling and condensing

We are learning how to:
- change the state of a substance from liquid to gas and back to liquid
- explain that these are reversible changes
- classify these changes of state as physical changes.

Boiling and condensing 》》》

When a liquid is heated it will eventually reach its **boiling** point and become a gas or **vapour**. When a gas is cooled it turns back into a liquid.

If you leave a bottle containing a cold drink out on a table, it soon gets wet on the outside. The air near the bottle, which contains water vapour, has been cooled by the bottle. This causes the vapour to condense, forming liquid drops of water on the outside of the bottle.

Activity 2.9.1

Investigating boiling and condensation

Is the change from liquid to gas a change of substance?

Here is what you need:
- Bunsen burner
- Conical flask
- Distillation apparatus.

Here is what you should do:
1. Set up the apparatus as shown in Fig 2.9.2.
2. Place some water in the flask that is above the Bunsen burner.
3. Heat the flask of water until the water boils, and observe what happens.
4. What happens in the conical flask?
5. After a few minutes, turn off the Bunsen burner and allow the apparatus to cool down. Then, dip a piece of cobalt chloride paper into the liquid in the conical flask. Does it turn pink? What does that show?
6. Is this what you expected? Has the water changed?

FIG 2.9.1 Condensation deposits droplets of water on the outside of these cold bottles

As water is heated, the increase in temperature causes the particles to gain energy and move further apart. As the particles move faster and further apart, they overcome the bonds between them and the bonds eventually break.

Water becomes a gas when the bonds between its particles no longer exist. This is called boiling. The change from liquid to gas can be described as **vaporisation**.

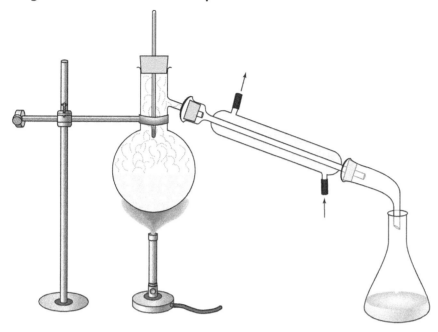

FIG 2.9.2 Apparatus set up for boiling and condensing

When you cool a gas, heat energy flows from it. The loss of heat energy causes the particles to slow down, and the bonds between them begin to take effect. The particles move closer together. The gas changes to a liquid. This is **condensing**. Another word to describe this change is **liquefaction**.

Boiling and condensing are reversible, depending on the direction of flow of heat energy.

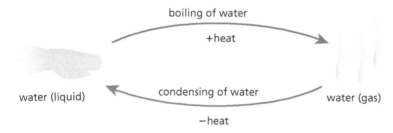

FIG 2.9.3 Heat energy flow, boiling and condensing

The change from liquid to gas and back did not change the substance of the water. This means that these changes of state are both physical changes.

Check your understanding

1. Draw a flow chart to show what happens to boiling water leaving a kettle and condensing on cold tiles behind a hob. Make sure you use the words 'boil', 'temperature', 'condense' and 'heat energy'.

Key terms

boiling the process by which a liquid changes to a gas at its boiling point

vapour another term to describe a gas

vaporisation when a substances changes into a gas

condensing when a gas changes to a liquid

liquefaction when either a solid or a gas changes to a liquid

Evaporation

Evaporation »

If we leave a glass of water on a sunny window ledge the water will soon 'disappear' leaving an empty glass. The water changes to gas and is lost to the atmosphere.

Although the sun's rays may be hot they will certainly not heat the water to 100 °C so the water in the glass does not boil. Clearly there must be another process by which a liquid can change into a gas.

The particles in a liquid are in continual motion and have different amounts of kinetic energy. At any given time a very small proportion of particles at the surface of the liquid have sufficient energy to move away from the rest and become gas. This process is called **evaporation**.

FIG 2.10.1 Evaporation takes place at any temperature

Evaporation is the process by which a liquid becomes a gas or vapour at temperatures below its boiling point.

The term '**vapour**' is often used to describe a gas below the boiling point of the liquid from which it has formed. Thus when water evaporates it may be said to form water vapour rather than steam.

There are two important differences between evaporation and boiling:

• A substance only boils at a particular temperature, called its boiling point. Evaporation takes place at any temperature but is greatest in warm moving air.

• Evaporation only takes place at the surface of a liquid while boiling takes place throughout the liquid.

Although evaporation takes place below the boiling point of a liquid, energy is still needs to convert a liquid to a vapour. Therefore when a liquid evaporates, it absorbs energy.

Activity 2.10.1

Showing that evaporation requires energy

Here is what you need:

- A few drops of a volatile liquid such as rubbing alcohol (isopropyl alcohol), ethanol or ethoxyethane.

Here is what you should do:

1. Place a few drops of a volatile liquid on the back of your hand.
2. Hold your hand still and allow the liquid to evaporate.
3. What sensation do you feel on your skin?
4. What does this tell you about evaporation?

Fun fact
....

Evaporation and condensation are very common processes. They occur in the water cycle.

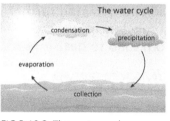

FIG 2.10.3 The water cycle

Volatile liquids are liquids which have low boiling points and evaporate very quickly.

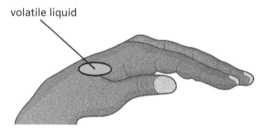

FIG 2.10.2 Cooling effect of evaporation

If a small amount of a volatile liquid, such as ethoxyethane, is placed on the back of the hand, it will evaporate so quickly that the place where it was will feel cold. The heat energy needed for evaporation is taken from the skin.

Check your understanding

1. Fig 2.10.4 shows a simple way of keeping soda cool on a warm day. A bottle of soda is wrapped in a wet towel or damp newspaper and placed in the breeze.

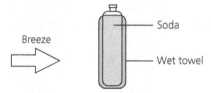

FIG 2.10.4

Explain how it works.

2. Another way to cool a drink is to bury it in the sand at a river when the students are fishing or shrimping. Explain how this cools the drink.

Key terms
....

evaporation the process by which a liquid changes to a gaseous state below its boiling point

vapour another term for a gas

volatile quickly changing from liquid to gas at room temperature

Sublimation

We are learning how to:
- define sublimation and desublimation
- explain why sublimation and desublimation are physical changes.

Sublimation »

We know that when heated, a solid can change to a liquid, and when heated further, a liquid can change to a gas. Do all substances go through the liquid state when they change from solid to gas?

Activity 2.11.1

Observing sublimation

Here is what you should do:
1. Place some crystals of iodine in a test tube.
2. Warm the test tube gently in a hot water bath.
3. Observe the iodine vapour that rises from the crystals.
4. Cool the test tube in iced water.
5. Discuss these questions:
   a) What does heat do to iodine?
   b) Is any liquid present?
   c) Where do the fumes go?
   d) What happens to the fumes? Why does this happen?
   e) Can you tell whether the change is a physical one?

FIG 2.11.1 Iodine being heated

Fun fact

FIG 2.11.3

Because solid carbon dioxide sublimes at room temperature, it is often used at concerts or in plays where the effect of fog or smoke is required. The cold carbon dioxide condenses the water vapour in the air and creates clouds of water droplets.

Some substances change directly from their solid state to their gaseous state when heat energy is applied. As soon as the heat energy is withdrawn, they return to the solid state. The processes are referred to as **sublimation** and **deposition**.

Sublimation and deposition are both physical changes – the substance, such as the iodine in Activity 2.11.1, remains unchanged. Another example of a solid that sublimes is **dry ice** (Fig 2.11.2).

FIG 2.11.2 Dry ice subliming

Dry ice is the solid form of carbon dioxide. It is used as a cooling agent. Dry ice has two advantages in that its temperature is lower than that of water ice and it leaves no residue. It is used for preserving frozen foods such as ice cream. The extreme cold of dry ice makes it dangerous to handle. Direct contact with dry ice can cause severe burns.

FIG 2.11.4 Moth balls are made of naphthalene, which sublimes at room temperature. The naphthalene vapour that is given off is harmful to insects, so it can keep clothes free from insect pests

Key terms

sublimation the change from a solid to a gas without liquefaction

deposition the change from a gas to a solid without liquefaction

dry ice the solid form of carbon dioxide

Check your understanding

1. Is heating needed for sublimation?

2. Draw a diagram, like those in Figs 2.8.3 and 2.9.2, to show sublimation and deposition of a substance.

Heat and expansion

We are learning how to:

- describe what happens to the volume of solids, liquids and gases when they are heated
- explain some of the results of expansion.

Heat transfer >>>

Metals are particularly good at transferring **heat**.

If we place a metal spoon in a hot drink the end of the spoon soon becomes hot. The heat from the drink is transferred through the spoon.

Activity 2.12.1

Observing how heat travels along a metal rod

Here is what you need:

- Metal rod
- Paper clips × 4
- Candle wax
- Support such as a clamp
- Heat source
- Stopwatch or clock that counts in seconds.

Here is what you should do:

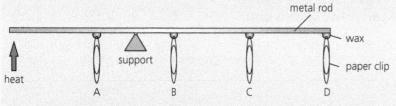

FIG 2.12.2

FIG 2.12.1

1. Light a candle and use molten wax to attach four paper clips (A, B, C and D) at equal distances from each other along a metal rod (see Fig 2.12.2). Don't attach a paper clip to one end of the rod as that is where the rod will be heated.

2. Measure the distance from each blob of wax to the end of the rod that will be heated.

3. Support the rod using a stand and clamp or similar apparatus.

4. Start heating the end of the rod and start timing.

5. Record the number of seconds before each paper clip falls off.

> **Fun fact**
>
> In the early days of railway building, engineers didn't realise how much metal railway lines would expand on hot days. The lines buckled and the trains were derailed.

6. Record your results in a table like the one below.

Paper clip	Distance from heat source / cm	Time taken to fall off / s
A		
B		
C		
D		

TABLE 2.12.1

7. Describe and explain any pattern you see in your results.

Expansion

When metal objects are heated they don't just get hot, they also **expand**.

In Fig 2.12.3, the ball is too big to pass through the hoop when the hoop is cold. When the hoop is heated it expands and the ball can pass through.

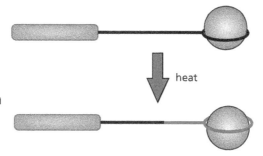

heat

FIG 2.12.3 Metals expand when heated

FIG 2.12.4 Large metal structures like bridges expand in hot weather. One end of the bridge is fixed while the other sits on rollers so the bridge can expand without buckling

FIG 2.12.5 Railway lines expand in hot weather so a small gap is left between sections of line to allow for expansion

FIG 2.12.6 Overhead power lines are always erected with a small amount of sag to allow for contraction in cold weather

Check your understanding

1. The bar shown in Fig 2.12.7 fits exactly into the holder when they are both at room temperature.

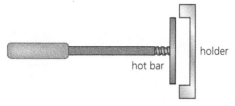

holder

hot bar

FIG 2.12.7

Predict whether the bar would still fit if:

a) The bar was heated but the holder remained at room temperature.

b) The bar remained at room temperature but the holder was heated.

c) Both the bar and the holder were heated.

d) The bar remained at room temperature and the holder was left overnight in a freezer at a temperature of −20 °C.

Key terms
....

heat form of energy

expand get bigger

63

Diffusion

We are learning how to:

- explain how gases fill the space available
- describe some of the results of diffusion.

Diffusion ⟫⟫

You already know something about **diffusion** although you may never have heard it called by this name.

Sometimes when you get home from school you know what you are having for supper before you see it because you can smell it! Tiny food particles travel in the air from the kitchen to the rest of your home. This spreading out of particles is called diffusion.

FIG 2.13.1 Something smells good!

Activity 2.13.1

Investigating diffusion

Your teacher will help you with this activity.
Here is what you need:

- Shallow open container like a watch glass
- Volatile liquid that has a characteristic smell, like a perfume
- Stopclock with a second hand.

Here is what you should do:

1. Students sit or stand in rows at different distances from the teacher's table.

2. The teacher will pour a small amount of perfume onto a watch glass on the table at the front of the class and start the clock.

3. Each student should raise his or her arm and note the time when he or she first smells the perfume.

4. Which students were first to smell the perfume?

5. Which students were last to smell the perfume?

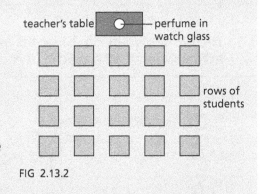

FIG 2.13.2

In Activity 2.13.1 the perfume spreads out, or diffuses, from the watch glass to all parts of the classroom.

Substances can diffuse in water just like they do in air, but more slowly. To observe diffusion in water, you can use a crystal of a coloured compound, like potassium manganate(VII), that is soluble in water.

Substances diffuse more quickly in hot water than in cold water. If you want to increase the rate of diffusion you can stir the mixture. This is why people use a spoon to stir sugar into their tea and coffee.

Investigating whether diffusion takes place more quickly in hot water than cold water

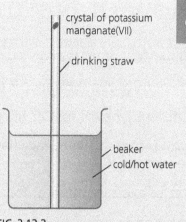

2.13

Here is what you need:

- Large beaker half filled with water at room temperature
- Large beaker half filled with hot water at about 60 °C
- Wide drinking straw
- Two small crystals of potassium manganate(VII).
- A timer or stopwatch.

Students will work in groups for this activity; some members of the group will work with the hot water and some with the cold.

FIG 2.13.3

1. Place the beakers of water about 15 cm from each other so you can watch them both at the same time.
2. Put the drinking straw into the cold water so one end is resting on bottom of the beaker at its centre.
3. Carefully drop a small crystal of potassium manganate(VII) down the straw so it lands at the bottom of the beaker and then remove the straw to allow the potassium manganate to disperse.
4. Carry out steps 2 and 3 with the beaker containing hot water.
5. Observe how the purple colour of potassium manganate(VII) diffuses through the water.
6. Compare how quickly the diffusion takes place in the cold water and hot water by comparing the beakers every five minutes. Photograph the beakers together if you have a camera or a cell phone that can take pictures.
7. In which solution did diffusion take place quicker?

Check your understanding

1. Fig 2.13.4 shows what happened when a crystal of potassium manganate(VII) was placed in a beaker of water and left for 24 hours.

   Explain why the appearance of the water changed.

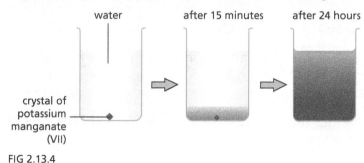

FIG 2.13.4

Fun fact

Diffusion takes place through solids but this happens much more slowly than through liquids and gases.

Key term

diffusion the spreading out of particles of a substance to fill the space available

Review of Nature of matter

- All matter has mass and volume.

- The density of a substance is the mass of a particular volume of that substance.

- Different types of matter have different densities.

- Substances that feel heavy have a high density while substances that feel light have a low density.

- Matter exists in different forms, called states.

- States of matter can be determined by physical properties:
  - Solids have definite volume and shape.
  - Liquids have definite volume but no definite shape.
  - Gases have neither definite volume nor shape.

- All matter is composed of tiny particles held together by bonds.

- The particles in any substance remain the same. However, both the energy of the substance and the arrangement of its particles can change. These two changes determine whether a substance is solid, liquid or gas.

- The particles in a solid only do not have sufficient energy to overcome the strong forces of attraction between them. They are held in fixed positions close to each other. Solids have a definite shape; don't flow, so they cannot be poured; and cannot easily be compressed.

- The particles in a liquid have sufficient energy to move about but cannot escape from each other. They are close together but can change position. Liquids take the shape of their container; flow, so they can be poured; but cannot easily be compressed.

- The particles in a gas are very energetic and move about very quickly. They are spread out much more than solids and liquids. Gases take the shape of their container; flow, so they can be poured; and can easily be compressed because of the space between the particles.

- When the temperature of a solid is increased, its particles gain energy. The solid may change to a liquid and then to a gas as the particles move faster and further apart.

- When the temperature is decreased, the processes are reversed.

- Melting, boiling, condensing and freezing are all physical changes, since they do not change the chemical composition of the substances. All these processes occur in nature, for example in the water cycle.

- When a liquid evaporates it changes from a liquid to a vapour at temperatures below its boiling point. Evaporation is greatest in warm, moving air.

- Boiling occurs at a particular temperature and takes place throughout a liquid. Evaporation can occur at any temperature but only takes place from the surface of a liquid.

- Some substances sublime as they change directly from solid to gas. Deposition is the reverse process. Both sublimation and deposition are physical changes.

- Dry ice is the solid form of carbon dioxide, used as a cooling and a preserving agent. It sublimes, so is also used to create 'fog' effects.

- Substances expand when they are heated.

- During the construction of structures like bridges, railway lines and power lines, allowances are made for the expansion and contraction that takes place with changes in temperature.

- Diffusion is the spreading out of particles to fill the space available.

- Diffusion occurs most rapidly in gases, more slowly in liquids and very slowly in solids.

Review questions on Nature of matter

Questions

Select the correct responses to the following questions.

1. Which of the following is NOT a gas?

    a) Air   **b)** Car exhaust fumes   **c)** Fluffy cotton balls

2. Is an uninflated balloon heavier or lighter than when it is blown up?

    a) Heavier   **b)** Lighter   **c)** The same

3. When a sponge under water is squeezed what escapes?

    a) Soapy water   **b)** Particles of sponge   **c)** Air from the spaces in the sponge

4. Which material is a solid?

    a) Honey   **b)** Oil at room temperature   **c)** Cotton wool

5. To change metal from solid to liquid, what must you do?

    a) Heat it   **b)** Cool it   **c)** Bend it

Indicate whether the following are true or false.

6. All liquids keep their volume.

7. Solids can be cut since their bonds are weak.

8. Liquid cannot keep a shape.

9. The bonds in gases keep them from expanding in volume.

10. A solid will expand to fill its container.

11. Gases are easy to pour from one container to another.

Answer the following questions.

12. How is matter defined?

13. a) Identify the states in which matter exists.
    b) Give two characteristics of each state.
    c) Illustrate the arrangement of particles in each state.

14. a) Do the particles in a given substance differ?
    b) Name the two factors that determine the state of a substance.

15. Identify the similarity between the following:

    a) Solids and liquids   **b)** Liquids and gases.

16. From your experiments, give one factor that is responsible for change of state.

17. Explain what happens to the particles in:

a) solids as they change to liquids

b) gases as they change to liquids.

18. Identify two uses of dry ice. For each use, state the property of the dry ice that allows it to have this use.

19. Fig 2.RQ.1 shows the changes from solid to liguid to gas and back. What are the numbered processes occuring on each arrow?

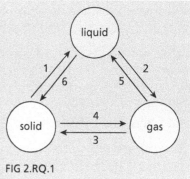

FIG 2.RQ.1

20. **a)** State two differences between evaporation and boiling.

b) Explain why:

i) Wet clothing is hung out on a line to dry and not left in a heap.

ii) Wet clothing dries quicker on warm days.

21. **a)** Why is it misleading to say that lead is heavier than aluminium?

b) What do we mean when we say that lead in denser than aluminium?

22. Explain the following observations.

a) When a railway is laid small gaps are left between the lengths of steel rails.

b) One side of a steel bridge is fixed and the other side sits on rollers.

c) A key fits into a lock when it is cold but not after it has been heated.

d) Power lines which are attached to pylons in the summer are not pulled tight but allowed to sag.

23. **a)** Why can petrol often be smelt on the forecourt of a filling station?

b) Why would a person be foolish to smoke a cigarette on the forecourt even if they were stood away from the pumps?

24. **a)** Why do people stir their tea after they have put sugar into it?

b) If left unstirred would a spoon of sugar diffuse more quickly in a cup of hot coffee or a glass of cold fruit juice?

Measuring the freezing points of vegetable oils

Ms Dinnal bought a bottle of coconut oil to use in her cooking. She placed it in her refrigerator overnight. The next morning she was surprised to find that it had become solid and she could no longer pour it from the bottle.

If the manufacturer had given the freezing point of coconut oil on the bottle this problem might have been avoided. Information about the freezing points of different vegetable oils would help people to decide the best ways of storing them.

FIG 2.SIP.1 Oil becomes solid when it is cool

1. You are going to work in a group of 3 or 4 to investigate the freezing points of different oils commonly used in food preparation. The tasks are:

   - To find out what vegetable oils are available from your local stores.
   - To devise a method of measuring the freezing points of vegetable oils.
   - To measure the freezing points of different vegetable oils.
   - To ensure the values you obtain are accurate.
   - To prepare an information sheet which provides useful advice about the storage of different vegetable oils.

   a) Have a look at the different vegetable oils available in your local shops. Here are some you might see:

olive oil	soybean oil	corn oil
coconut oil	canola oil	vegetable oil

   Are all of these available? Are there any other vegetable oils not on this list?

   You will need to obtain a small quantity of each of the oils available for the investigation you are going to carry out.

   b) How are you going to find the freezing point of a vegetable oil? This is when it changes from liquid vegetable oil to solid vegetable oil or fat.

   For any particular vegetable oil the freezing point of the liquid oil is the same as the melting point of the solid fat. Perhaps it would be easier to find the temperature at which the fat melts?

   All of the oils mentioned above, in their fat states, have melting points in the range −20 °C to 25 °C.

FIG 2.SIP.2 Freezer and fridge temperatures

A typical refrigerator cools to 5 °C and a typical freezer cools to –20 °C. Does this give you any idea how you might convert a liquid oil to a solid fat?

How are you going to measure the melting point of a fat? It isn't easy to insert a thermometer into the fat. Perhaps you could allow a liquid oil to solidify into a fat with a thermometer already inserted in it?

One method of finding the melting point of a solid oil is to gently warm a test tube of the solid in water at room temperature. You might be able to suggest a different way.

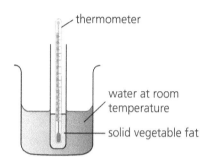

FIG 2.SIP.3 One way of measuring melting point

c) A good scientific method should be repeatable and provide consistent results.

How often will you repeat your procedure for each vegetable oil to ensure your answers are as accurate as possible?

d) Construct an information sheet that can be used to guide people about the storage of vegetable oils. You might simply give a list of vegetable oils and their melting points or you might be more ambitious and include some illustrations and advice on storage.

e) Give an oral presentation to the class on what you found out about the freezing points of different vegetable oils. This should include a demonstration of how you obtained your data.

Unit 3: Cells and organisms

We are learning how to:

- find out more about living things
- look at how living things are different and how they are similar
- identify features they all have that tell us they are living.

Things can be similar and different ⟫

Animals and plants come in different shapes and sizes. Some are very large, while others are so small they can only be seen with the help of a microscope.

Stones also come in different shapes and sizes. Does this mean that stones are living things? What are the features of living things that make them different from non-living things?

FIG 3.1.1 Living things

FIG 3.1.2 Non-living things

All living things exhibit all seven characteristics, but some characteristics are easier to observe than others in different organisms.

Seven characteristics are associated with living things:

Growth

Respiration

Irritability

Movement

Nutrition

Excretion

Reproduction

If you take the first letter of each characteristic, it spells GRIMNER. This will help you to remember them.

Living things are made of cells

All living things are composed of cells. Some simple organisms, like *Euglena*, consist of only a single cell. More complex organisms contain many millions of cells.

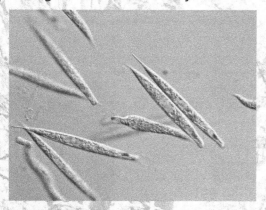

FIG 3.1.3 *Euglena*, a unicellular organism

The human body contains many different types of cell. Tissues are composed of many similar cells. The different tissues combine to form organs.

Photosynthesis

FIG 3.1.4 Green plants carry out photosynthesis

Many of the cells in a green plant contain a pigment called chlorophyll. These cells carry out a process called photosynthesis, which makes food for the plants and also for the animals that eat the plants.

Substances enter and leave cells

In order for organisms to flourish, substances must be able to pass into and out of cells. Substances do this by the processes of diffusion and osmosis.

Key terms

growth the process by which an organism gets bigger

respiration the process by which living things obtain energy

irritability the ability of an organism to respond to a stimulus

movement the ability of an organism to move position or place

nutrition the process by which an organism obtains food necessary to sustain its life

excretion the removal of waste products from the body of an organism

reproduction the process by which organisms produce offspring

Growth

We are learning how to:

- describe the characteristics of living things
- relate growth to living things.

Growth >>>

All living things **grow** in size as they get older. Some organisms grow to their **mature** size very quickly, while others may take many years.

Redwood trees may live and grow for a thousand years. They are the largest and tallest trees in the world.

The appearance of many plants and animals does not change much as they grow. The mature organism often looks to be just a large version of the **young** one.

In some animals, we can see small differences between young and mature forms. For example, a calf looks very similar to a cow but is much smaller, its features are more rounded and it does not have horns.

FIG 3.2.1 Redwood trees

FIG 3.2.2 Cow and calf

Activity 3.2.1

Growing red bean seedlings

Here is what you need:

- Mung bean or red bean seeds × 5 • Water. • Ruler

1. You will be measuring the growth of bean seedlings but first you will need to germinate the seeds. Design a seed box 12″ × 12″ × 4–6″ deep to be used in planting your seeds.

2. Select five seedlings, plant them, and label them A, B, C, D and E.

3. Measure the height of each seedling every day for 10 days. You can expect some readings of 0. Put your measurements into a table.

4. Calculate the average height of the seedlings each day, by adding all five seedling heights for one day together and dividing the total by 5.

5. Use your data to plot a graph of average height of the seedlings against time. What do you notice from your graph? Answer these questions:
   - Are the all the heights of the seedlings the same?
   - Do the seedlings increase in height each day?
   - Is the increase the same for all the seedlings each day?

Certain groups of animals, including some insects and amphibians, completely change their appearance as they grow. This process is called **metamorphism**.

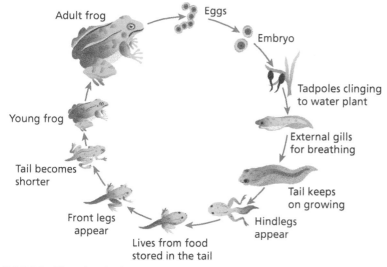

FIG 3.2.3 Life cycle of a frog

When frogspawn hatches, the eggs develop into tadpoles. The tadpoles have no limbs and they have gills for breathing in water. As the tadpoles grow, they slowly change into adult frogs with four limbs. The gills are replaced by lungs, because the frog spends most of its adult life on land.

Fun fact

Different parts of the human body grow at different rates, so our shape changes as we grow up. Fig 3.2.5 shows how the relative sizes of the different parts of the body change as a baby grows into an adult.

FIG 3.2.5 Shape of the human body at different ages

Check your understanding

1. Fig 3.2.4 shows how the average height of boys and girls increases with age.

   a) During which period is the rate of growth greatest?

   b) At what age does a girl reach her maximum height?

   c) At what age does a boy reach his maximum height?

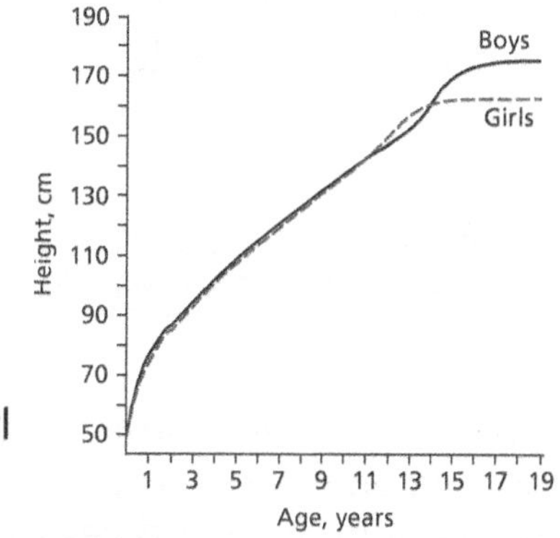

2. Are frogs the only animals that carry out metamorphosis? Use the internet to carry out research on animals that carry out metamorphosis.

FIG 3.2.4

Key terms

growth the process by which an organism gets bigger

mature an organism in the latter part of the life cycle when it may be fully grown

young an organism in the early part of the life cycle when it is still growing

metamorphism a process that brings about a complete change in the form of an animal as part of its life cycle

Respiration

We are learning how to:
- describe the characteristics of living things
- relate respiration to living things.

Respiration »»

Respiration is the process by which cells release energy. Respiration is not the same as breathing.

All the cells in an organism require energy for the different chemical processes going on inside them. This energy is obtained from the reaction between **glucose** and **oxygen**.

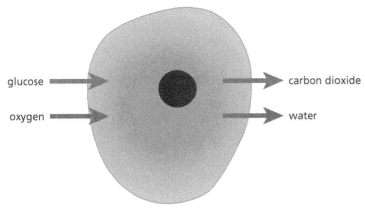

glucose ⟶

oxygen ⟶

⟶ carbon dioxide

⟶ water

FIG 3.3.1 Cell respiration

Glucose and oxygen enter the cell, while **carbon dioxide** and **water** – the waste products of respiration – pass in the opposite direction.

The process of cell respiration can be summarised by the following word equation:

glucose + oxygen ⟶ carbon dioxide + water + energy

All organisms carry out cell respiration. The way in which an organism obtains the glucose and oxygen it needs and expels the waste products formed depends on its structure.

Many simple unicellular organisms, like amoeba, live in water. Substances can pass into the organism from the water and also out of the organism.

FIG 3.3.2 An amoeba

FIG 3.3.3 Banana leaf

Green plants make their own glucose by photosynthesis. Gases pass into and out of a plant through openings in the underside of the leaves.

Complex animals like people have too many cells for substances to be able to pass directly into cells from outside the body. Such animals have a circulatory system that carries materials to and from cells. In people, glucose and oxygen are carried to the cells in the blood. Waste products are removed from cells in the same way.

FIG 3.3.4 Humans have a circulatory system that allows our cells to respire

The amount of energy a person needs each day depends on two factors, which are their:

- basic metabolic rate
- level of activity.

Basic metabolic rate is the energy needed to keep the body working even when it is resting. This accounts for 60–70% of the body's energy needs.

Check your understanding

1. Copy and complete the following sentences.

   a) Respiration is the process by which living things produce ....

   b) The substances needed for cell respiration are .... and ....

   c) The waste products of cell respiration are .... and ....

Key terms

respiration the process by which living things obtain energy

glucose a simple sugar

oxygen a gas found in the air, which is made during photosynthesis and used up during respiration and combustion

carbon dioxide a gas found in the air, which is made during respiration and combustion, and used up during photosynthesis

water a liquid required by all forms of life to survive

Irritability

We are learning how to:

- describe the characteristics of living things
- relate irritability to living things.

FIG 3.4.1 *Daphnia* are attracted to the light in search of food

Irritability »»

Irritability is sometimes also called **sensitivity**. It is the ability of a living thing to react to a **stimulus**. This might be a chemical stimulus such as a harmful chemical, or it might be a physical stimulus such as light.

Microscopic aquatic organisms like algae are attracted to light. The organisms that feed on them, like *Daphnia*, also respond to this stimulus.

Activity 3.4.1

Measuring the response of woodlice to light and damp

Here is what you need:

- Flat box
- Filter paper × 2
- Piece of card, at least half the size of the box
- A dozen or so woodlice
- Soil.

Here is what you should do:

1. Draw pencil lines to divide a flat box into four regions.
2. Sprinkle a thin layer of soil on the bottom of the box. This will help the woodlice to move about.
3. Place pieces of damp filter paper in opposite corners of the box.

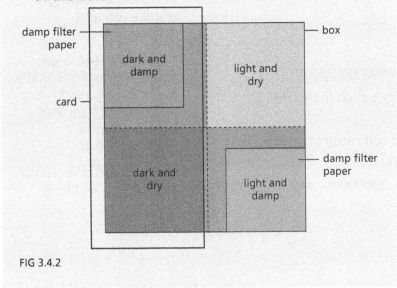

FIG 3.4.2

Key terms

irritability the ability of an organism to respond to a stimulus

sensitivity an awareness of different stimuli such as light or sound

stimulus a thing or an event that causes an organism to respond in a particular way

response the behaviour of an organism to a stimulus

4. Cover half of the box with a piece of card so that you create four regions where the conditions are different.
5. Place the woodlice in the box and leave the box somewhere quiet for 15 minutes or so.
6. After 15 minutes, count the number of woodlice in each of the four regions. How did the the woodlice react to the stimuli of light and damp?

Animals respond to stimuli in different ways. A stimulus might attract some animals while having the opposite effect on others. For example, woodlice prefer a dark environment, so they will respond to light in an appropriate way.

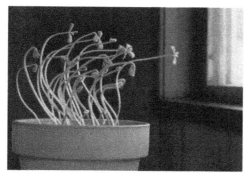

FIG 3.4.3 Plant stems grow towards the light

The **response** of plants to stimuli such as light is much less obvious that the response of animals, because it takes much longer. If a plant is placed on a windowsill, its stem will grow towards the light over a few days.

Activity 3.4.2

Video record of plant movement
1. Place some mung bean seeds on damp cotton wool in a petri dish and allow them to germinate.
2. Place the container with the germinated seedlings in a covered box with a hole on one side.
3. Leave the seedlings in the box in the laboratory or the classroom for a few days.
4. Set up your camera to record using the time lapse technique, ensure that it is stable.
5. Place the container with the seedlings, ensuring that it is in the view path of the camera. Use a lamp to shine light on the seedlings.
6. Record the changes with your camera.

Check your understanding

1 The human body has sense organs that respond to a number of different stimuli. Copy and complete Table 3.4.1 with the sense organ that responds to each of the stimuli given.

Fun fact

FIG 3.4.4 **a)** Charaille and **b)** the 'sensitive plant'

Some plants respond to touch. Charaille (*Momordica charantia*) has hairy tendrils that sense touch and can move away from the object touching it. The 'sensitive plant' (*Mimosa pudica*) has leaves that curl up when touched.

Stimulus	Sense organ
Touch	
Sound	
Taste	
Light	
Smell	

TABLE 3.4.1

Movement

We are learning how to:

- describe the characteristics of living things
- relate movement to living things.

Movement >>>

All organisms are able to **move**, although animals move much more freely than plants. Plants tend to move their parts while staying in a fixed position.

The way in which animals move depends on their body structure.

Birds are able to move through the air because their upper limbs have evolved into **wings**. They also have other features, like hollow bones, that help them to fly.

FIG 3.5.1 Birds can fly because they have wings

FIG 3.5.2 Some land animals move about on legs

Many land animals move about on two, four or more **legs**.

Animals like snakes are able to move about on land and in water without the help of legs. They push themselves along by contractions of their muscular bodies.

FIG 3.5.3 Some land animals are able to move without legs

FIG 3.5.4 Fish are ideally shaped to move through water

Fish have muscular bodies that can propel them through the water. They have evolved without limbs, which would spoil their streamlined shape and slow them down.

FIG 3.5.5 Mangroves can alter the directions in which their leaves point

Although plants cannot change their position, they can turn their leaves either towards the Sun, in order to get the most light, or away from the Sun to protect them from burning during the hottest part of the day.

Check your understanding

1. Penguins are birds but they cannot fly. Instead, they spend a lot of time feeding on fish in the water.

FIG 3.5.6

Look carefully at Fig 3.5.6 and describe three features that help the penguin to move easily through water.

Key terms

movement the ability of an organism to move position or place

wings structures that some animals have, which enable them to fly

legs structures that some animals have, which enable them to move about on the ground

Nutrition

We are learning how to:

- describe the characteristics of living things
- relate nutrition to living things.

Nutrition »»

Nutrition is about obtaining food. Food is needed for producing energy and for growth. Plants and animals have completely different ways of obtaining their food.

FIG 3.6.1 Plants make their own food

Plants make their own food by a process called **photosynthesis**. In this process, simple substances are built into complex substances. Energy from sunlight is used to convert carbon dioxide and water into glucose, as shown in the word equation below.

carbon dioxide + water + energy $\longrightarrow$ glucose + oxygen

You will learn more about photosynthesis later in the book.

All animals depend either directly or indirectly on green plants for their food. Animals obtain their food by eating plants or by eating animals which eat plants.

Animals can be placed in groups according to whether they eat only plants, only other animals or a mixture of plants and animals.

Herbivores eat only plants. Cattle and goats are examples of herbivores. The mouths of herbivores have few or no front teeth, but they have large back teeth so they can grind the plant material before swallowing it. Insects like grasshoppers are also herbivores. They eat the foliage of plants.

FIG 3.6.2 Goats are herbivores

FIG 3.6.3 Grasshoppers are herbivores

FIG 3.6.4 Mongooses are carnivores

FIG 3.6.5 Praying mantises are carnivores

FIG 3.6.6 Vultures eat the bodies of dead animals

Carnivores eat only other animals. In order to kill other animals, carnivores like the mongoose have sharp front teeth that can pierce and cut into flesh. Some insects are carnivores. The praying mantis eats other insects.

Scavengers like vultures are another group of carnivores. They do not actually kill other animals but feed off the bodies of animals that are already dead.

Omnivores are animals that eat both plants and other animals. Rats are omnivores. They eat fruits, seeds and roots but they will also eat small animals like worms and insects.

After animals have eaten, complex substances are broken down into simple substances.

Activity 3.6.1

Creating a green drink

Some green plants have leaves are a food source for humans. These plants include spinach, callaloo, celery and many others.

1. Design an interview sheet and use this to interview the teachers of the Food and Nutrition department of your school in order to discover the range of plants whose leaves or stems are eaten by humans.
2. Create a green drink using the leaves of one of these plants.

Check your understanding

1. Table 3.6.1 lists what some different animals eat. Decide whether each animal is a herbivore, a carnivore or an omnivore.

Name of animal	What the animal eats
Deer	Plant material
Owl	Small rodents like rats
Bat	Insects
Mouse	Seeds
Monkey	Fruit and insects

TABLE 3.6.1

Fun fact

Most people are omnivores. They have a diet of foods obtained both from plants and from animals, although they may not eat meat at every meal. Some people are herbivores. They eat foods obtained from plants but will not eat foods obtained from animals. These people are called vegetarians.

Key terms

nutrition the process by which an organism obtains food necessary to sustain its life

photosynthesis the process in plants by which carbon dioxide and water are converted to glucose using energy from sunlight

herbivore an animal that eats only plants

carnivore an animal that eats only other animals

omnivore an animal that eats both plants and other animals

Excretion

We are learning how to:

- describe the characteristics of living things
- relate excretion to living things.

Excretion 》》

'**Metabolism**' is a collective word to describe all the chemical processes that occur within a living organism in order to maintain life. Many of these processes produce waste products that are harmful to the organism and must be removed. **Excretion** is the removal of these waste products.

In simple unicellular organisms that live in water, like *Paramecium*, waste products are able to pass directly out of the organism into the surrounding water.

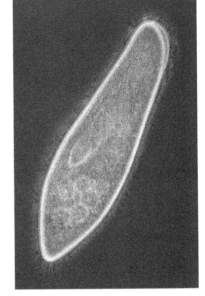

FIG 3.7.1 *Paramecium*

Activity 3.7.1

Waste products of yeast metabolism

Here is what you need:

- Dried yeast
- Test tubes × 2
- Rubber bung
- Limewater
- Teaspoon
- Sugar
- Test-tube stand
- Glass tubing
- Distilled water.

Here is what you should do:

1. Place half a teaspoon of dried yeast and half a teaspoon of sugar in the bottom of a test tube.

2. Add water until the test tube is about one-third full.

3. Half fill a second test tube with limewater.

4. Connect the test tubes with glass tubing as shown in Fig 3.7.2.

5. Leave the apparatus to stand for a day or until bubbles of gas from the solution of sugar and yeast start passing through the limewater.

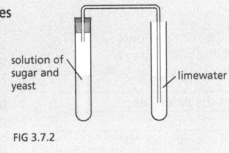

solution of sugar and yeast

limewater

FIG 3.7.2

6. How does the appearance of the limewater change?

7. Which gas is a waste product of yeast metabolism?

8. Carefully remove the bung from the test tube containing the solution of sugar and yeast. Smell the contents of the tube. Can you identify another waste product of yeast metabolism?

Activity 3.7.2

Uses of yeast

1. Working in groups, carry out research on the uses of yeast, in the laboratory, in the home and in industry.

2. Use this information to construct charts/posters to display the information that you have collected.

Some of the carbon dioxide and water produced by green plants as a result of cellular respiration is used in photosynthesis. Remaining unwanted gases are able to pass out through tiny pores in the underside of leaves.

FIG 3.7.3 Plant resin

Plant metabolism also creates other waste products that are not gases. These are removed from the plant as resin or sap that passes out through the plant stem.

Complex organisms like people have an excretory system that removes and expels waste products. Carbon dioxide and water are expelled from the body each time we breathe out. Other waste products are removed from the blood by the kidneys and lost from the body in solution as urine.

Excretion should not be confused with the removal of undigested food from the body as faeces. This process is not excretion, because the faeces are not the waste products of metabolic processes; it is called **egestion**.

> **Fun fact**
>
> Plants excrete some waste materials in their leaves. The materials are removed when the old leaves fall off the tree to be replaced by new leaves.

Check your understanding

1. Mangroves grow in soil that contains a high level of salt. In order for metabolic processes to take place, the mangrove must remove some of the salt it absorbs. The salt is lost in solution through pores in the underside of the leaves. When the water evaporates, salt crystals may be formed.

FIG 3.7.4 Salt crystals on the underside of a mangrove leaf

Is this an example of excretion? Explain your answer.

Key terms

metabolism the chemical processes that take place within an organism in order to maintain life

excretion removal of waste products from the body of an organism

egestion the expulsion of undigested food from the body

Reproduction

We are learning how to:

- describe the characteristics of living things
- relate reproduction to living things.

All living things reproduce, otherwise organisms would become extinct very quickly. They need to make more individuals of the same kind. However, there is a great variety in the frequency of **reproduction** and the mechanisms by which organisms reproduce. Some organisms reproduce sexually and some reproduce asexually.

Sexual reproduction

Sexual reproduction involves two 'parent' organisms. Flowering plants are able to carry out sexual reproduction. The flower is the reproductive organ.

A hibiscus flower has both **male** parts and **female** parts. During sexual reproduction, the male sex cells, or pollen, are transferred from one flower to the female parts of another flower. This process is called **pollination** and is carried out by insects. The pollen then combines with female sex cells to eventually produce seeds.

Sexual reproduction is the way in which many animals reproduce. In some animals, like birds, the young develop outside the body of the parents. In other animals, like people, the young develop inside the body of the female.

FIG 3.8.1 Hibiscus flower

FIG 3.8.2 Banana sucker

FIG 3.8.3 In some animals the male and the female look similar, but in others they look different

FIG 3.8.4 Air plant, also known as life plant or miracle leaf is a member of the *Bryophyllum* group of plants

Asexual reproduction

Single-celled organisms are able to reproduce by a process called binary fission. Initially, the nucleus divides and then the cytoplasm around it divides. The result is two new cells.

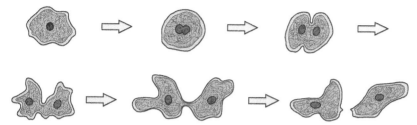

FIG 3.8.5 Binary fission

Binary fission is also the process by which complex organisms like people produce new cells for growth and repair.

In **binary fission**, only one 'parent' organism is involved, therefore it is an example of **asexual reproduction**.

Plants are able to undergo asexual reproduction (sometimes also called **vegetative reproduction**) in a number of ways.

New banana plants grow from suckers that develop on the root of a single banana plant. Once the sucker is big enough, it can be separated from the parent plant and will grow on its own.

Members of the *Bryophyllum* group of plants are able to grow new plantlets at the ends of their leaves. These plantlets eventually fall off and start to grow in the ground.

Check your understanding

1. Strawberry plants can reproduce by sending out runners. At the end of each runner a new plant forms.

FIG 3.8.7

   a) Is this an example of asexual or sexual reproduction?

   b) Explain your answer to a).

   c) What name is sometimes given to this process when it occurs in plants?

Fun fact

Slugs, snails and earthworms have both male and female reproductive organs.

FIG 3.8.6 Earthworms are hermaphrodites

Animals that have both male and female sex organs are called hermaphrodites.

Key terms

reproduction the process by which organisms produce offspring

male the sex that produces sperm (animal) or pollen (plant)

female the sex that produces ovum (animal) or ovule (plant)

pollination the transfer of pollen from one flower to another either on the same plant or a different plant

binary fission this describes a cell dividing into two equal parts

asexual reproduction a method of reproduction which involves only one parent

vegetative reproduction asexual reproduction in plants

What are cells?

We are learning how to:

- compare plant and animal cells according to their structure and function
- describe cells.

What are cells? 〉〉〉

If you look carefully at Fig 3.9.1 you will see that, although the building is very large, it is built of many individual stones or bricks. If you imagine that an organism is like a building, then, just as the building is composed of stones or bricks, so the organism is composed of **cells**.

The cell is the building block from which all organisms are composed. All plants and animals contain large numbers of cells. Scientists estimate that the human body contains many millions of cells.

A cell is very tiny and cannot be seen with the unaided eye. In order to see a cell, we need a **microscope**. You will have an opportunity to prepare slides and look at cells using a microscope in later lessons.

All cells have certain common features. There are some differences between animal cells and plant cells. You will learn about the differences later in the course. For now we will look at some different examples of each type of cell.

FIG 3.9.1 Killarney mansion on Queen's Park Savannah, Port of Spain

Animal cells

Animals are composed of cells. Look carefully at these different examples of animal cells.

Although these are all animal cells, they have different shapes and appearances because they have different jobs to do in the body. For example:

- Cheek cells protect the inside of the cheek.
- Red blood cells travel through the blood, carrying oxygen to all the other cells of the body.
- Nerve cells carry nerve impulses.

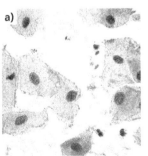
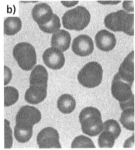
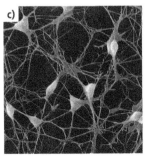

FIG 3.9.2 Different animal cells:
a) cheek cells
b) red blood cells and
c) nerve cells

Plant cells

Plants are also composed of cells. Look carefully at the following pictures of cells in a plant.

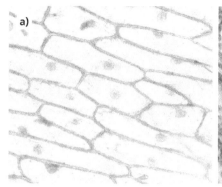 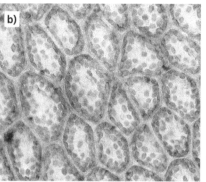 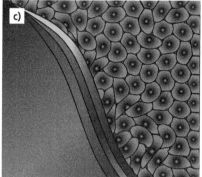

FIG 3.9.3 Different plant cells, from the: **a)** onion epidermis, **b)** leaf and **c)** stem

Once again, we can see a difference in the shape and form of the different cells.

- Epidermal cells are rectangular and fit together to protect the fleshly leaf.

- Leaf cells contain chlorophyll, which captures energy from sunlight. Green plants use this energy in the process of photosynthesis, by which they make food.

- Stem cells are closely packed to provide the plant with support.

Check your understanding

1 Fig 3.9.5 shows the great Khufu pyramid at Giza in Egypt.

FIG 3.9.5 The great Khufu pyramid at Giza

a) From what is the pyramid made?

b) How is the structure of the pyramid similar to that of a plant or an animal?

Fun fact

Australian scientists have discovered fossils in rocks that tell us that cells emerged on Earth at least 3.4 billion years ago.

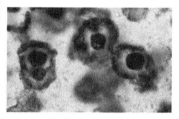

FIG 3.9.4 Fossil evidence of cells

Key terms

cell the building block from which all living things are made

microscope an optical device that magnifies an image of a specimen, so more detail can be seen

Using a microscope

We are learning how to:

- compare plant and animal cells according to their structure and function
- use a microscope.

Using a microscope >>>

A **microscope** is an instrument that makes objects look bigger than they really are.

A simple microscope has two lenses, called the **objective lens** and the **eye lens** or eyepiece. These two lenses are fixed at each end of a tube. The tube can be moved up and down in order to focus on an object.

The power, or magnification, of a lens is usually given as a number. For example, an eye lens might typically have a magnification of times 10, or ×10. An objective lens might have a magnification of times 4, or ×4.

The overall magnification of a microscope is the product of the magnification of the two lenses. In the example above, the overall magnification of the microscope is 10 × 4 = ×40. This means that everything you see under the microscope appears 40 times bigger than it actually is.

More sophisticated microscopes have three or four objective lenses of different powers on a rotating turret. A low-power lens can be used to locate the area of a specimen to be examined, and then the high-power lens is used to see that part of the specimen in more detail.

Microscope slides and cover slips

Microscope **slides** are rectangles of glass. The normal size is 75 mm by 25 mm and about 1 mm thick. Microscope slides may be flat or have a small cavity ground into the surface.

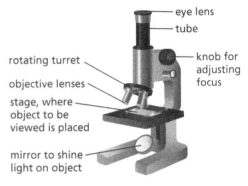

eye lens
tube
knob for adjusting focus
rotating turret
objective lenses
stage, where object to be viewed is placed
mirror to shine light on object

FIG 3.10.1 A microscope with three different objective lenses

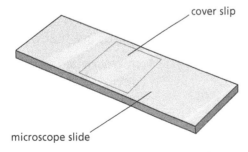

cover slip

microscope slide

FIG 3.10.2 A microscope slide and cover slip

Fun fact

Around 1590, two Dutch spectacle makers, Hans and Zacharias Janssen, started experimenting by putting several lenses into tubes. They discovered that looking at an object through the tube made it appear much larger than when looked at with a single magnifying glass.

The Janssens' work was taken up by another Dutchman, Antonie van Leeuwenhoek. Van Leeuwenhoek made a working microscope that allowed him to see things that no one had ever seen before, like bacteria, blood cells and tiny organisms in droplets of pond water.

A specimen is placed on a microscope slide and covered by a much thinner piece of glass called a **cover slip**. The cover slip prevents the specimen from drying out and also protects the objective lens of the microscope.

Activity 3.10.1

Learning how to use a microscope

Microscopes are expensive to buy and to repair if they are damaged. In this activity you will have the opportunity to learn how to use a microscope correctly.

Here is what you need:

- Microscope
- Microscope slide.

Here is what you should do:

1. Look at the microscope and identify the parts.

2. Turn the knob that adjusts the focus, and observe how this moves the tube up and down.

3. Put a small specimen in the middle of the microscope slide. For example, you might want to look at a hair.

4. Adjust the microscope until the object comes into focus.

5. If your microscope has more than one objective lens, look at the object through low power first and then through high power.

6. Practise drawing what you can see.

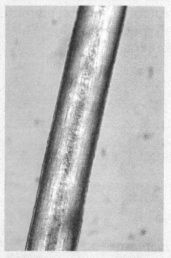

FIG 3.10.3 A human hair as seen under a microscope magnified 25 times

Key terms

microscope an optical device that magnifies an image of a specimen, so more detail can be seen

objective lens the lens on a microscope that is nearer to the specimen being observed

eye lens the lens on a microscope that is nearer to the eye

slide a small oblong piece of glass upon which a specimen to be examined under a microscope is placed

cover slip a thin disc or square of glass placed on top of a specimen to be viewed under a microscope

Check your understanding

1. A microscope has an eyepiece of power ×10 and three objective lenses of power ×4, ×10 and ×40.

   a) What is the lowest power at which you could look at a specimen with this microscope?

   b) What is the highest power at which you could look at a specimen with this microscope?

Preparing a specimen of animal cells

We are learning how to:

- compare plant and animal cells according to their structure and function
- prepare animal cells to view with a microscope.

Different types of animal cells ▶▶

There are many different kinds of animal cell. We are going to look at four examples, and observe slides of animal cells under a microscope.

Cheek cells

Cheek cells are obtained by gently scraping the inside of the cheek.

Cells prepared for examination under a microscope are often stained with a dye so the structure can be seen more clearly. **Cheek cells** are stained with **methylene blue**. This makes the cells look blue when examined.

Key terms

cheek cell cells from the inside of the cheek in the mouth

methylene blue a dye that stains materials blue

skeletal muscle cells cells from the large muscles of the body that are long fibres which can move

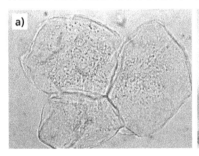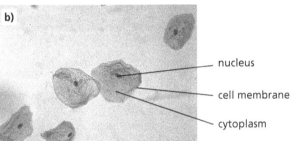

a)   b)

nucleus

cell membrane

cytoplasm

FIG 3.11.1 Cheek cells: **a)** not stained and **b)** stained

It is possible to see the nucleus, cytoplasm and cell membrane of cheek cells after staining.

Muscle cells

There are different types of muscle cells in the body. The muscles we use to move our arms and legs are called **skeletal muscle**.

The muscle cells are long and thin. Do they remind you of the structure of meat? If a person chooses to eat meat, this meat will be mostly animal muscle.

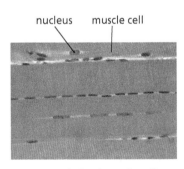

nucleus   muscle cell

FIG 3.11.2 Skeletal muscle cells

Liver cells

The liver has many important functions in the body. **Liver cells** are much rounder than muscle cells.

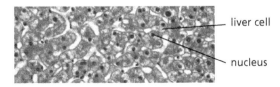

liver cell

nucleus

FIG 3.11.3 Liver cells

Blood cells

Blood consists of a mixture of particles, including **blood cells**, suspended in a watery liquid called plasma.

There are white blood cells and red blood cells. You will learn more about their different functions later in your course.

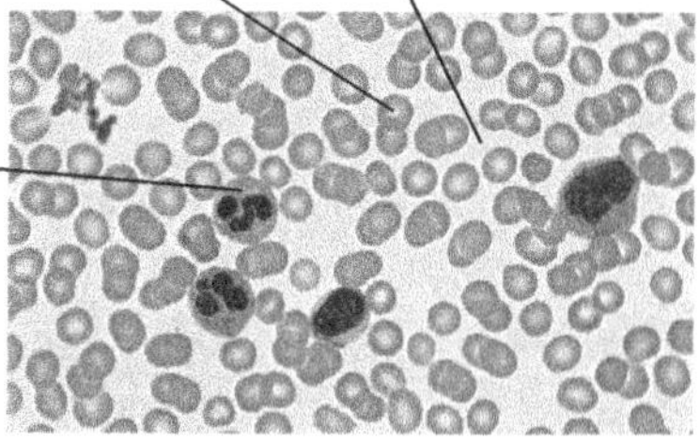

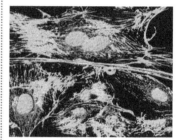

FIG 3.11.4 Blood cells

Activity 3.11.1

Observing prepared slides of animal cells

Here is what you need:

- Microscope
- Prepared slides of animal cells.

Here is what you should do:

1. Place the first prepared slide under the microscope.

2. Start by observing it under low power. This makes it much easier to see the arrangement of cells, and you can identify particular groups of cells that you would like to see in greater detail.

3. When you have identified a part of your specimen that you would like to see in greater detail, move the slide so that this part is in the centre of the image.

4. Change the combination of eyepiece and objective lens to see the image under higher power.

5. Draw the cells and label any parts you recognise.

6. Repeat this for the other prepared slides.

Fun fact

Cells are sometimes stained with combinations of different dyes.

FIG 3.11.5 Different parts of a cell show up in different colours

Each dye is absorbed by a different part of the cell, so the parts are easier to see.

Check your understanding

1. Fig 3.11.6 shows some skin cells.

   Draw a single skin cell and label the nucleus, cytoplasm and cell membrane.

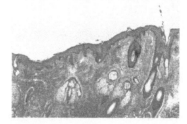

FIG 3.11.6 Skin cells

liver cells cells from the liver that are round and carry out many functions

blood cells cells from the blood; some are red and carry gases, others are white and fight infections

Structure of an animal cell

We are learning how to:

- compare plant and animal cells according to their structure and function
- identify parts of an animal cell.

Structure of an animal cell »

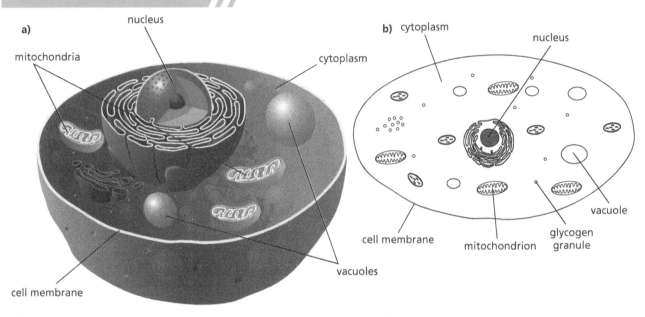

FIG 3.12.1 Structure of an animal cell **a)** in three dimensions and **b)** as a scientific drawing

Fig 3.12.1 shows the structure of a typical animal cell. The cell has three main parts.

1. The **nucleus** is a small dark structure inside the cell. It contains **chromosomes**.

2. The **cytoplasm** is a jelly-like substance that fills the cell. It contains these important structures:
   - **Mitochondria** (singular mitochondrion)
   - **Glycogen grains**
   - Small **vacuoles**.

3. The **cell membrane** is a thin layer that surrounds the cell.

Nucleus

The nucleus controls and coordinates all the processes, such as respiration, that take place within the cell. If you imagine a cell to be like a computer, then the nucleus is the central processor. The nucleus was the first part of a cell to be discovered by scientists.

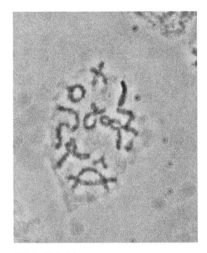

FIG 3.12.2 Chromosomes are found in the nucleus of a cell

Chromosomes in the nucleus are important because they contain all the information the cell needs to function and to replicate itself – the genetic information.

Cytoplasm

The cytoplasm is where many complex chemical reactions take place in the cell. Cytoplasm has a number of important components, including the following:

- **Mitochondria**
  These are sometimes the described as the cell's power plant, because they provide the cell with energy.

- **Glycogen grains**
  All cells need glucose to carry out cell respiration. Glucose is stored inside cells in the form of another chemical, called glycogen. When the cell needs energy, glycogen molecules can be quickly converted back into glucose.

- **Small vacuoles**
  Small vacuoles are often found in animal cells. They are bubbles filled with water containing dissolved substances.

Cell membrane

The cell membrane surrounds the cell. It has an important function in controlling the movement of materials into and out of the cell.

Check your understanding

1. Fig 3.12.3 represents an animal cell.

   a) What are the parts labelled X, Y and Z?

   b) Which of the labelled parts:

   i) contains mitochondria?

   ii) controls the activity of the cell?

   iii) controls the movement of substances into and out of the cell?

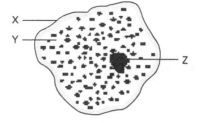

FIG 3.12.3

Fun fact

The first cells were observed by Robert Hooke in 1665 using a crude microscope of low magnification. One observation was from very thin slices of cork. He saw many tiny pores that we now know are filled with air. Robert Hooke did not know what the pores were, but called them 'cells', from the Latin word *cella*, which means a small room like the ones where monks lived.

Key terms

nucleus the part of a cell that controls the processes that take place in it

chromosomes structures in the nucleus that carry information which allows the cell to duplicate itself

cytoplasm a jelly-like substance that surrounds the nucleus in a cell

mitochondria structures in the cytoplasm responsible for the release of energy

glycogen grains the way in which glucose is stored in animal cells

vacuoles a space within a cell containing fluid

cell membrane a membrane that surrounds the cell and controls what enters and leaves it

Preparing a specimen of plant cells

We are learning how to:

- compare plant and animal cells according to their structure and function
- prepare plant cells to view with a microscope.

Preparing a specimen of plant cells »

Although there are many different kinds of plant cell, a convenient source of cells is the thin membrane of **epidermal cells** between the layers of an onion. This membrane is only one cell thick.

Onion cells are normally stained with **iodine solution** to make them easier to observe.

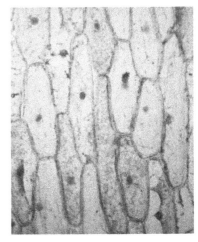

FIG 3.13.1 Epidermal cells in the membrane between the layers of an onion

Activity 3.13.1

Preparing and observing onion cells

Here is what you need:

- Microscope
- Cover slip
- Piece of onion
- Scalpel
- Scissors
- Microscope slide
- Iodine solution
- Tissue paper
- Tweezers
- Distilled water.

Here is what you should do:

1. Peel part of a layer from an onion. If you look carefully at the inside of the piece you have removed, you will see a thin membrane. This is duller in appearance than the outside of the onion.

2. Using a scalpel or small knife very carefully, start to separate the membrane from the onion, and then remove it completely using a pair of tweezers.

3. The piece of material you remove is likely to be far bigger than will sensibly fit on a microscope slide. Using scissors, cut off a piece of the membrane, about 5 mm by 5 mm, and place it on a clean slide.

4. Add a drop of distilled water to make a wet mount.

5. Add a single drop of iodine solution to the distilled water and leave it for 1 minute.

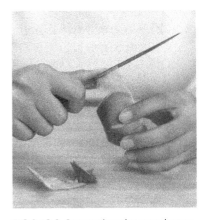

FIG 3.13.2 Separating the membrane from the ion with a knife. Be careful!

6. After this, lower a cover slip onto the specimen so that it is sitting on the surface of the water.

7. Using a tissue as a wick, remove the excess water and iodine solution from under the cover slip. The tissue might discolour when it comes into contact with the iodine.

8. Ensure your slide is dry before placing it on the microscope. Wipe the top and bottom of the slide with a clean tissue. Take care not to move the specimen and cover slip.

9. Onion cells are significantly larger than cheek cells, so you will not need such high magnification to see them in detail. Set your microscope to the lowest power available and look at your specimen.

10. Find an area of your specimen where the onion cells have stained well and you can see lots of detail. Make this the centre of your field of view and change your microscope to a higher power.

11. Draw your onion cells.

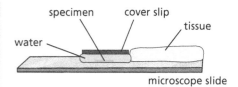

FIG 3.13.3 Preparing the microscope slide

Onion cells are regular in shape. They are rectangular and fit together like bricks in a wall. Onion cells have both thin cell membranes and thick cell walls, so the outline of the cell is easy to see.

Although onion cells are plant cells, they do not have chloroplasts. Chloroplasts contain the green pigment chlorophyll. The plant uses chlorophyll during photosynthesis to make food (glucose).

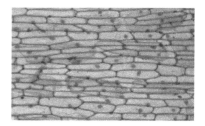

FIG 3.13.4 The iodine stains the onion cells so they can be seen clearly

Check your understanding

1. a) Explain why specimens to be viewed with a microscope are stained with a dye.

   b) Suggest a suitable dye with which to stain onion skin cells.

2. Fig 3.13.4 shows onion cells seen through a microscope. The eyepiece used was ×10 and the objective lens ×4. The distance across the picture is 32 mm.

   a) What was the overall magnification of the microscope?

   b) Estimate the length of an onion cell in millimetres (mm).

Key terms

epidermal cells cells that come from the epidermis or skin between layers

iodine solution a substance that forms a distinctive blue-black colour when mixed with starch

Structure of plant cells

We are learning how to:

- compare plant and animal cells according to their structure and function
- identify parts of a plant cell.

Structure of plant cells

If you look carefully at the structure of a typical plant cell shown in Fig 3.14.1, you will see that it has some parts that are found in animal cells and also some extra parts that are not.

Plant cells contain a nucleus, chromosomes, cytoplasm, mitochondria and a cell membrane. There are also some important additional features.

1. Plant cells generally have much larger vacuoles than animal cells do.

2. All plant cells are surrounded by a **cell wall** made of cellulose.

3. Many plant cells have chloroplasts. Inside the **chloroplasts** is a green pigment called chlorophyll. This pigment is responsible for the colour of the leaves and stems of plants.

4. Plant cells contain **starch grains**.

Cell wall

A plant cell has both a cell membrane and, surrounding it, a cell wall. Cell walls are made of cellulose. The cell wall around a plant cell has a different role from the cell membrane. Tough and rigid, but also with some flexibility, the cell wall provides a plant cell with structural support while protecting it.

The cell wall also prevents the plant cell from over-expanding when it absorbs water. Together, the rigid cell wall and the pressure of water in the cells allow a plant stem to stand upright.

Chloroplasts

Chloroplasts are the most obvious feature of a plant cell. The word 'chloroplast' comes from Greek words meaning 'green thing'. Green is the colour we associate with all plants.

The role of chloroplasts is to trap energy from sunlight during photosynthesis. Each chloroplast contains a pigment called chlorophyll, which is responsible for the characteristic green colour.

a)

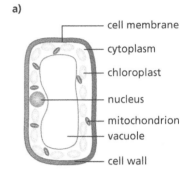

- cell membrane
- cytoplasm
- chloroplast
- nucleus
- mitochondrion
- vacuole
- cell wall

b)

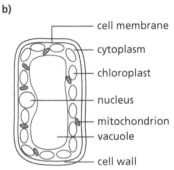

- cell membrane
- cytoplasm
- chloroplast
- nucleus
- mitochondrion
- vacuole
- cell wall

FIG 3.14.1 Structure of a plant cell a) in colour b) as a scientific diagram

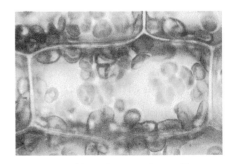

FIG 3.14.2 Chloroplasts in plant cells

Vacuoles

Vacuoles are present in many animal cells, but they are generally much larger in plant cells. Vacuoles are filled with an aqueous solution containing different chemical compounds.

The role of vacuoles varies greatly according to the type of cell. In plant cells, the role of the large vacuole includes:

- isolating substances that might be harmful

- containing waste products

- transferring unwanted substances from the cell

- maintaining turgor pressure within the cell, which allows the plant to support structures like leaves and flowers.

Starch grains

Glucose is not stored in plant cells as glycogen, but as starch. Glycogen and starch are examples of storage polymers.

Activity 3.14.1

Modelling a plant cell

Here is what you need:

- Modelling clay – several different colours

- Modelling tools to flatten and cut the clay

- Toothpicks

- Small labels.

Here is what you should do:

1. Build a model of a plant cell using modelling clay.

2. Your model should have all the parts that you would expect to see in a plant cell. Each part should be a different colour.

3. When you have completed your model, write a label for each part. Attach each label to one end of a toothpick and place the other end in the part.

Check your understanding

1. Fig 3.14.4 shows the structure of a plant root cell.

   a) Name parts A, B, C and D.

   b) What structure found in a leaf cell would you not expect to find in a root cell?

   c) Explain your answer to **b)**.

Fun fact

Algae (singular alga) were once considered to be single-celled plants.

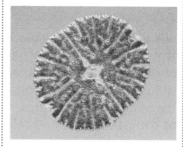

FIG 3.14.3 An alga

Algae are no longer classified as plants, because they have no root, stem or leaves.

Key terms

cell wall the outer layer of a plant cell

chloroplast the structures in a plant cell that contain the green pigment chlorophyll, which traps the energy needed for photosynthesis

starch grain the way in which glucose is stored in plant cells

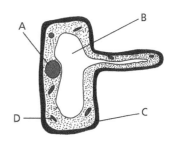

FIG 3.14.4 Plant root cell

Cells, tissues and organs

We are learning how to:

- recognise the relationships between specialised cells, tissues, organs and organ systems
- describe cells, tissues and organs.

Cells, tissues and organs >>

Cells are the building blocks from which all **organisms** are formed. Some simple organisms, like yeast and amoeba, are described as unicellular because they are composed of a single cell. More complex organisms, like humans and flowering plants, are described as multicellular because they are composed of many cells.

These cells are arranged in **tissues**, the tissues form parts of an **organ**, and the organ forms part of an organ **system**.

cells → tissues → organs → systems → organism

Fig 3.15.1 shows the relationship between cells, tissues and organs in the human digestive system.

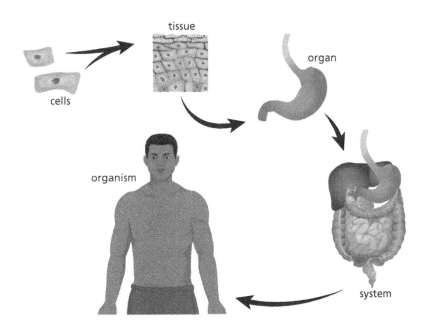

FIG 3.15.1 The human digestive system

Activity 3.15.1

Organ systems in the human body

1. Write a song or poem to show the relationship between the cells, tissues and organs of any of the organ systems in the human body.

The human digestive system contains different types of cell. These cells are grouped together to form different types of tissue.

The stomach is an organ of the digestive system. It is composed of different types of tissue. The small intestine is another organ of the digestive system. It is also composed of tissue, but the tissue is different to that of the stomach.

The stomach, small intestine and other organs together are called the digestive system.

The digestive system is one of a number of different organ systems in the human body.

The transport system of a plant moves water and nutrients from one part to another. It contains different types of cell grouped to form different types of tissue. The root is an organ of the transport system.

The transport system is one of a number of different organ systems in a plant.

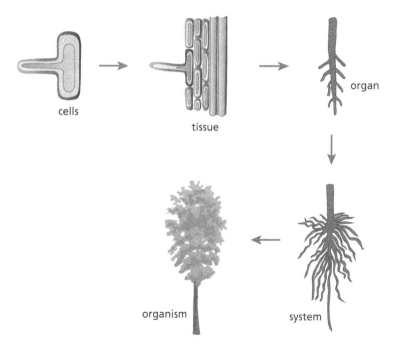

FIG 3.15.2 The transport system of a plant

Key terms

cell the building block from which all living things are made

organism a living thing

tissue a collection of cells that are similar in structure

organ a part of an organism which is self-contained and has a particular function

system a collection of two or more organs that have related functions

Check your understanding

1. Arrange the following in order, starting with the simplest:

   system   tissue   cell   organism   tissue

Unicellular organisms

We are learning how to:

- recognise the relationships between specialised cells, tissues, organs and organ systems
- identify unicellular organisms.

Multicellular and unicellular organisms 》》

Living organisms like human beings are described as **multicellular** because they consist of many millions of cells. Some very simple organisms consist of only one cell and are called **unicellular**. A unicellular organism carries out all of the activities of a living thing.

You have already seen one unicellular organism, *Euglena*, on page 73. We are going to look at some more examples.

Amoeba

Amoeba is a microscopic unicellular organism that lives in ponds and streams.

Amoeba has all of the characteristics of a living organism (see lesson 3.1). *Amoeba* can:

- carry out respiration to obtain energy
- absorb nutrients through its cell membrane
- excrete waste products through its cell membrane
- grow bigger
- reproduce by dividing into two new organisms
- move by allowing its cytoplasm to flow
- respond to stimuli, such as chemicals dissolved in water.

Paramecium

Paramecium is another microscopic unicellular organism that lives in water.

Paramecium also exhibits all of the seven characteristics of a living organism. It is able to move more quickly than *Amoeba*, because it is covered in tiny hair-like cilia that can beat together and propel it through the water.

Yeast

Yeast is a unicellular fungus. It has been used by people since ancient times to make bread and also to make alcoholic drinks by a process called fermentation.

Key terms

multicellular describes an organism that consist of many cells

unicellular describes an organism that consists of a single cell

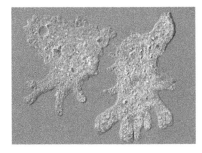

FIG 3.16.1 *Amoeba* is a unicellular organism

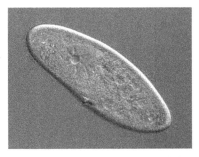

FIG 3.16.2 *Paramecium* is a unicellular organism

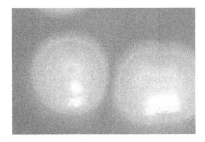

FIG 3.16.3 Every cell of yeast is a unicellular organism

Yeast reproduce by a process called **budding**. New cells grow from existing cells. If you look carefully at Fig 3.16.3 you will see that some of the yeast cells appear to be growing tiny buds. Each new small cell grows and then separates from its parent cell.

Activity 3.16.1

Observing yeast cells

Here is what you need:

- Microscope
- Microscope slide
- Cover slip
- Tissue paper
- Pipette
- Bottle of solution containing yeast cells.

Here is what you should do:

1. Shake the solution of yeast cells and remove a small amount in a pipette.

2. Place two drops of the yeast solution on to a microscope slide (see Fig 3.16.4).

3. Carefully lower a cover slip onto the specimen.

4. Place a tissue next to the cover slip and gently draw off the excess water.

5. Place the slide onto the stage of the microscope.

6. Observe the yeast cells using a low magnification. Look for yeast cells that appear to be budding, then examine these cells using a higher magnification.

7. Draw some budding yeast cells.

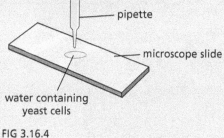

pipette

microscope slide

water containing yeast cells

FIG 3.16.4

Check your understanding

1. Fig 3.16.5 shows an organism called *Nassula*.

   a) Is *Nassula* better described as a unicellular organism or a multicellular organism?

   b) What characteristics of living organisms would you expect *Nassula* to have?

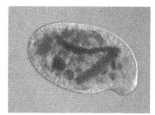

FIG 3.16.5 *Nassula*

> **Fun fact**
>
> At one time, the living world was classified by scientists into two kingdoms: plants and animals. Unicellular organisms were classified as either animals or plants, depending on which they most closely resembled. *Amoeba* was considered to be an animal while *Euglena* was considered to be a plant.
>
> Eventually scientists realised that unicellular organisms didn't really fit either of these kingdoms. The classification of the living world was revised to give the five kingdoms we have today. Unicellular organisms now have their own kingdom, called Protista.

Amoeba a tiny unicellular organism

Paramecium a tiny unicellular organism with small hairs on its surface

yeast a single-celled fungus which brings about fermentation

budding a system of reproduction where young cells bud off from the parent cell

Respiratory system

We are learning how to:

- recognise the relationships between specialised cells, tissues, organs and organ systems
- identify parts of the respiratory system.

Respiratory system ⟫

The **respiratory system** in the body is concerned with breathing. By breathing, the body can absorb oxygen from the air and release carbon dioxide. This takes place in the **lungs**. Oxygen is needed by the cells in the body for respiration.

We normally breathe air in through the nose. Air passes down the **trachea**, or windpipe, into the lungs. The trachea has rings of cartilage to prevent it from collapsing. You can feel the ridges if you put your thumb at the bottom of your throat just where your ribs start.

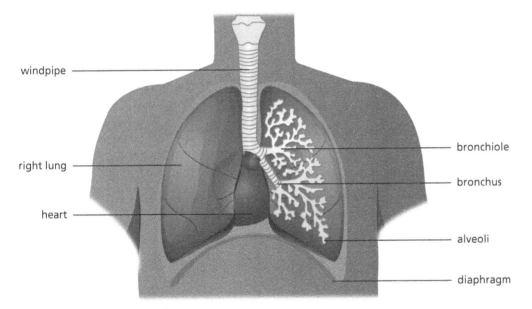

windpipe
right lung
heart
bronchiole
bronchus
alveoli
diaphragm

FIG 3.17.1 Respiratory system

The trachea divides into two bronchi (singular **bronchus**), one for each lung. Each bronchus further divides into many smaller tubes called **bronchioles**. At the end of each bronchiole is a cluster of tiny air sacs, or alveoli (singular **alveolus**).

The alveoli are well supplied with tiny blood vessels called capillaries. In these sacs, gases are exchanged. Carbon dioxide passes out of the blood into the lung, while oxygen passes from the lung into the blood.

Activity 3.17.1

Modelling the action of the lungs

Your teacher will assemble the apparatus needed for this activity.

Here is what you need:

- Bell jar
- Y-tube
- Balloons × 2
- Elastic band.

Here is what you should do:

1. Connect balloons to the two forks of a Y-tube using elastic bands.

2. Place the end of the Y-tube through a bung in the bell jar from the inside (see Fig 3.17.2).

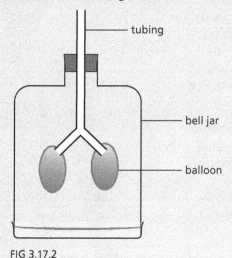

tubing

bell jar

balloon

FIG 3.17.2

3. Blow through the tube and observe what happens to the balloons.

4. Which part of the apparatus represents:

   a) the lungs?   **c)** the bronchi?

   b) the trachea?   **d)** the rib cage?

Check your understanding

1. Draw a flow diagram to show the order in which air passes through each of the following to reach the lungs:

   trachea   alveolus   bronchiole   nose   bronchus

Fun fact

The total surface area of the lungs is about the same as a tennis court! If all the tiny blood vessels in the lungs were unwound and added together, they would stretch for nearly 1,000 kilometres.

Key terms

respiratory system the system of the body that absorbs oxygen and excretes carbon dioxide and water

lung one of a pair of organs in which gases are exchanged

trachea the windpipe that carries air from the nose and mouth into and out of the lungs

bronchus one of two divisions at one end of the trachea that connects with a lung

bronchiole small airways formed by the repeated subdivision of a bronchus

alveolus an air sac found in groups at the end of a bronchiole

Circulatory system

We are learning how to:

- recognise the relationships between specialised cells, tissues, organs and organ systems
- identify parts of the circulatory system.

Circulatory system ⟩⟩

All the cells of the body need a continuous supply of nutrients and oxygen. These nutrients and oxygen dissolve in the blood, which is carried by the **circulatory system**.

The **heart** is the organ at the centre of the circulatory system. It pumps blood around the body. The **arteries** are thick blood vessels that carry blood away from the heart. Arteries are often shown in red on diagrams. The **veins** are thin-walled blood vessels that carry blood towards the heart. Veins are often shown in blue on diagrams.

The circulatory system has two parts – it is a 'double circulation' system. One part sends blood from the heart to the lungs, where the blood receives oxygen and is then returned to the heart. The other part sends blood from the heart to the rest of the body, where the oxygen is used, and the blood is then returned to the heart.

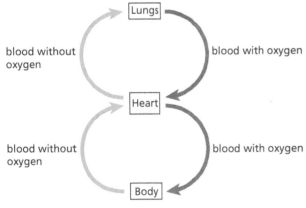

FIG 3.18.2 Double circulation

Taking a pulse

In order to pump blood around the body, the heart contracts and relaxes many times each minute. Each time the heart contracts, it forces a surge of blood along the arteries.

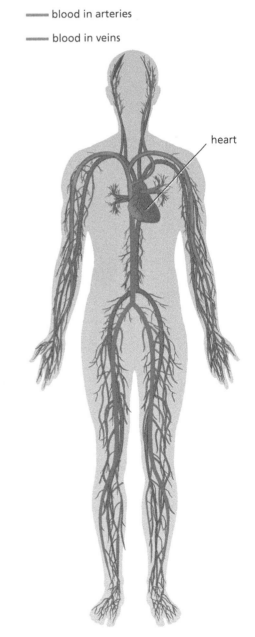

—— blood in arteries

—— blood in veins

heart

FIG 3.18.1 Circulatory system

If you put a fingertip on a point on the body where an artery crosses over a bone, you can feel the regular surge of blood. It is called a pulse. The pulse is most often taken on the wrist or neck.

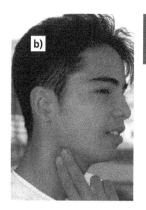

FIG 3.18.3 Feeling a pulse in the a) wrist and b) neck

An average adult has a pulse rate in the range of 60–80 beats per minute. This range of values is because people are all a little different.

Activity 3.18.1

Measuring a person's pulse rate

Before you can measure someone's pulse rate, you need to find that person's pulse. Here is what you need:

- Stopwatch.
1. Measure your partner's pulse rate by counting the number of pulses in 1 minute. Write the value in a table.
2. Repeat this two more times, so you have taken three measurements in all.
3. Collect the results of the students in the class. Use these results to create a comparison table of the average pulse rate of the boys against girls.
4. What was your average number of pulses in 1 minute over the three tests?
   - What do you notice?
   - Are all the pulse rates the same?
   - Are there noticeable trends?
   - Are the pulse rates of the boys similar to those of the girls?

Check your understanding

1. Which organ is at the centre of the circulatory system?
2. What is the name of the vessels that carry blood away from the heart?
3. What type of blood is pumped from the heart to the lungs?
4. What substances pass from the cells into the blood?

3.18

Fun fact

Blood travels around the body very quickly. When you are resting, blood takes about 1 minute to make a complete circuit of the body. When you are exercising, your body needs more nutrients and oxygen, so the blood circulates even faster.

Key terms

circulatory system the system that moves blood through the body in order to provide cells with nutrients and oxygen, and to remove waste products

heart the organ that pumps the blood around the circulatory system

artery a large blood vessel that carries blood away from the heart

vein a large blood vessel that carries blood towards the heart

Digestive system

We are learning how to:

- recognise the relationships between specialised cells, tissues, organs and organ systems
- identify parts of the digestive system.

Digestive system »»

The **digestive system** in the body obtains nutrients and water from the **foods** we eat.

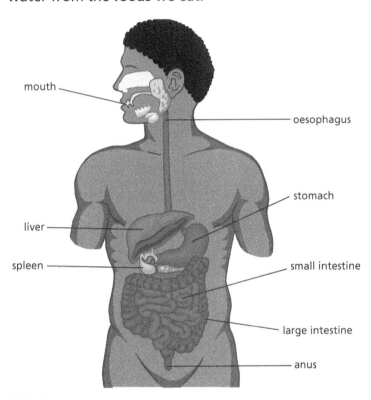

FIG 3.19.1 Digestive system

Food consists of complex chemicals that the body cannot absorb directly. During digestion, the food is physically broken down by the action of chewing in the **mouth**, and is chemically broken down by a group of substances called **enzymes**.

Digestion continues in the **stomach** and **small intestine**, where different types of food are broken down. Most of the nutrients produced by digestion are absorbed into the body through the walls of the small intestine. Some water is absorbed in the **large intestine**, and what remains passes out of the body as faeces.

Activity 3.19.1

Tracing the pathway of food through the alimentary canal

Here is what you need:

- Rectangular pieces of card, 5 cm × 3 cm, × 9.

Here is what you should do:

1. Make name cards for each of the following parts of the digestive system: large intestine, mouth, oesophagus, small intestine, stomach (see Fig 3.19.2a).

2. Make four arrow cards (see Fig 3.19.2b).

a) 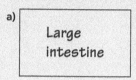   b)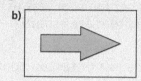

FIG 3.19.2 **a)** Name card and **b)** arrow card

3. Shuffle your cards and then place them on the table, face up.

4. Arrange the name cards and arrow cards to show the order in which food moves through the different parts of the digestive system.

5. Create a chart or poster to show the relationships between the different parts of the digestive system. Use presentation software or just use drawings or pictures to create your poster.

Check your understanding

1. Which part of the digestive system is an organ that is a muscular sac?

2. In which part of the digestive system:
   a) is most water absorbed?
   b) does digestion begin?
   c) does most absorption of nutrients take place?

3. What is the name of the group of substances that break down the complex chemicals in food into simpler substances that can be absorbed?

Fun fact

The body does not digest about 30% of the food that we eat. The undigested food passes through the digestive system and forms the bulk of the faeces.

Key terms

digestive system the system of the body that breaks down food and absorbs the nutrients which are released

food chemical substances that are broken down by the body to obtain nutrients and energy

mouth the first part of the digestive system where food is broken up and moistened, and where digestion starts

enzyme a chemical that helps to break down food

stomach an organ in the digestive system where food is held while digestion takes place

small intestine part of the digestive system where most absorption of nutrients takes place

large intestine part of the digestive system where water is absorbed

Excretory system

We are learning how to:

- recognise the relationships between specialised cells, tissues, organs and organ systems
- identify parts of the excretory system.

Excretory system ⟫

Excretion is the removal from the body of waste products produced by **metabolic processes**. The metabolic processes are those that occur inside the body and that are essential for life.

Note that excretion does not include the loss of faeces from the body. Faeces are not formed by metabolic processes, but are what is left after food has been digested. The correct term for the loss of faeces from the body is 'egestion'.

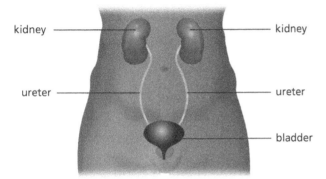

FIG 3.20.1 Excretory system

The main organs of the **excretory system** are the **kidneys**. The kidneys filter out soluble waste products from the blood and send them to the **bladder**, where they are stored and then expelled from the body as **urine**.

Excretion is not limited to the kidneys. The lungs also play a role in excretion. You have already seen how **carbon dioxide** gas passes from the blood into the lungs during breathing. Carbon dioxide is an excretory product formed during respiration.

The **skin** is also an important organ of excretion. Some of the water formed during respiration is lost through the skin during **sweating**. This water contains excess salt and urea (a waste product of the metabolism of proteins). Urea gives the body an unpleasant smell when we sweat.

> **Fun fact**
>
> People who have diseased or damaged kidneys can undergo a kidney transplant operation in which their kidneys are replaced by healthy kidneys, obtained from a person who has recently died. The first kidney transplant took place in 1950.

Excreting carbon dioxide

Here is what you need:

- Boiling tubes with bungs × 2
- Tubing
- T piece
- Limewater
- Stand and clamp × 2.

Here is what you should do:

1. Arrange the apparatus as shown in Fig 3.20.2. It is important that the tubes into the boiling tubes are the correct length and in the places shown.

2. Pour limewater into each boiling tube until it is about half full. Limewater turns cloudy in the presence of carbon dioxide gas.

3. Slowly breathe in and out through the mouthpiece.

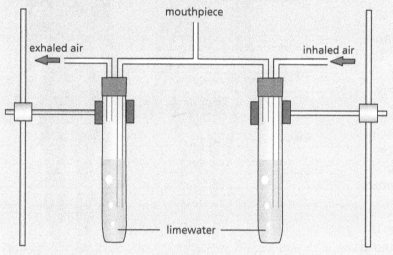

FIG 3.20.2

4. In which boiling tube does the limewater turn cloudy first?

5. What can you deduce about the concentration of carbon dioxide in exhaled air compared with that in inhaled air?

Check your understanding

1. What is excretion?

2. Give one example of a substance that the body excretes.

3. Why is the loss of faeces from the body not an example of excretion?

4. Name three organs involved with excretion.

Key terms

metabolic processes a process in the body that is essential for life

excretory system the system that removes the waste products of metabolism from the body

kidney one of a pair of organs that remove waste substances and some water from the blood

bladder a muscular sac that stores urine ready for excretion

urine a solution of waste substances, like urea and unwanted salts in water, that is expelled from the body at regular intervals

carbon dioxide a gas found in the air, which is made during respiration and combustion, and used up during photosynthesis

skin the outermost covering of the body, which is sensitive to certain stimuli and also has a function in excretion

sweat a solution of urea and salts in water, which is excreted through pores in the skin, especially during exercise or when the weather is warm

Skeletal and muscular systems

We are learning how to:

- recognise the relationships between specialised cells, tissues, organs and organ systems
- identify parts of the skeletal and muscular system.

Skeletal and muscular systems

The body is built around a framework of bones. These bones are the **skeletal system**, or **skeleton**. The skeleton contains 206 bones. The various muscles of the **muscular system** are attached to these bones.

The skeleton gives our body its shape and protects some of the organs. The skull protects the brain, while the ribcage protects the heart and lungs.

The skeleton also provides a framework of bones to which muscles attach. Muscles make it possible for the bones of the skeleton to move. Movement takes place at **joints** where bones meet.

For example, the action of muscles in the upper arm allows the arm to bend at the elbow joint. When one muscle relaxes and the other contracts, the arm bends at the elbow. When the roles of the muscles are reversed, the arm straightens.

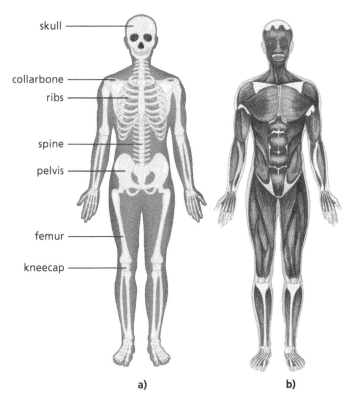

skull
collarbone
ribs
spine
pelvis
femur
kneecap

a)   b)

FIG 3.21.1 **a)** Skeletal system **b)** muscular system

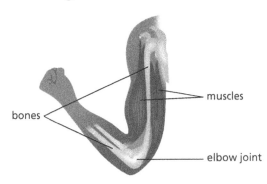

muscles
bones
elbow joint

FIG 3.21.2 Muscles make it possible for bones to move at joints

Activity 3.21.1

Modelling a hinge joint

A hinge joint is the type of joint you have at your elbow, at your knee and between the bones of your fingers.

Here is what you need:

- Craft sticks/wooden rods, washers and nuts
- Thin card
- Sticky tape.

Here is what you should do:

1. Use the craft materials provided to design a model hinge joint.

2. Here is an example using wooden rods and a card tube.

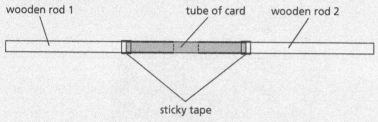

wooden rod 1   tube of card   wooden rod 2

sticky tape

FIG 3.21.3

3. Flatten the part of the cardboard tube between the wooden rods.

4. Bend the flattened part of the tube so the wooden rods come together.

5. How does the movement of the wooden rods resemble the movement of the bones in your arm?

6. Design an investigation to see if the length of the arms on your joint model affects their movement.

Key terms

skeletal system the system of the body, consisting of a framework of bones, that gives it shape

skeleton another name for the skeletal system

muscular system a collection of muscles connected to the skeleton, which makes movement possible

joint a place where two bones meet and can move relative to each other

Check your understanding

1. Fig 3.21.4 is an X-ray of the bones at a joint.

   a) In which part of the body do you think this joint is found?

   b) What makes it possible for the bones to move at a joint?

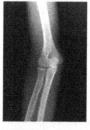

FIG 3.21.4

Reproductive system

We are learning how to:

- recognise the relationships between specialised cells, tissues, organs and organ systems
- identify parts of the reproductive system.

Reproductive system

The **reproductive system** is concerned with producing young. Unlike the other systems you have learned about, the parts of the reproductive system are different in males and females.

Male reproductive system

Key terms

reproductive system the part of the body involved in reproduction

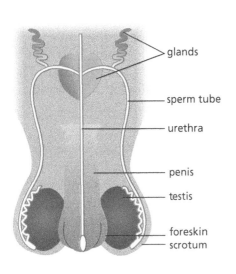

glands
sperm tube
urethra
penis
testis
foreskin
scrotum

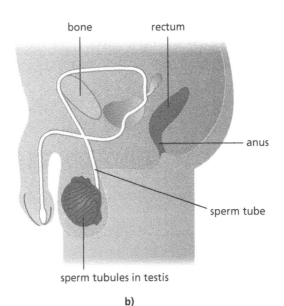

bone
rectum
anus
sperm tube
sperm tubules in testis

a)   b)

FIG 3.22.1 Male reproductive system from **a)** the front and **b)** the side

The sex organs in the male reproductive system are the testes (singular **testis**), which produce the sperm. The testes are held in a sack called the scrotum. The testes hang outside the body. Being outside the body keeps the testes below body temperature, which is necessary for sperm production.

Sperm are produced all the time and stored in tiny tubes outside the testes. The sperm eventually pass into the sperm tube and out of the body along the **urethra** and through the **penis**.

The glands have an important role in releasing a fluid that keeps the sperm alive when they pass out of the body. The mixture of sperm and fluid is called semen.

Fun fact

The human reproductive system functions only after we reach a certain age. For girls, this is usually between 10 and 15 years, and for boys between 11 and 16 years. The time when a person becomes sexually active is called puberty.

Female reproductive system

The female reproductive system is different from the male reproductive system, because it performs a different role.

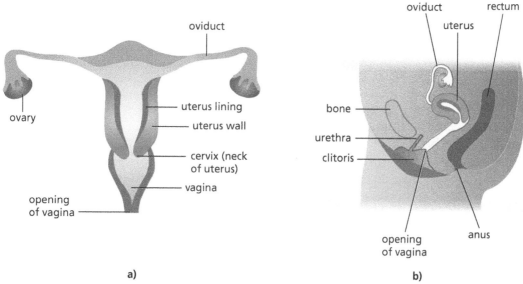

FIG 3.22.2 Female reproductive system from a) the front and b) the side

The **ovaries** are the female sex organs, and produce eggs. The ovaries are attached to the inside of the abdomen, just a little below the kidneys.

Once a girl becomes sexually mature, the ovaries release an egg approximately every 28 days. This is called ovulation. The egg passes into the funnel-shaped opening of the oviduct, and from there travels along the oviduct to the **uterus**, or womb.

If the egg is not fertilised, it will eventually pass out of the body through the cervix and **vagina** together with some of the thickened uterus wall. If the egg is fertilised, it will implant itself in the thickened wall of the uterus and develop into a baby over the following 9 months.

Check your understanding

1. Which parts of the male reproductive system are:

   a) outside the body?

   b) inside the body?

2. Which parts of the female reproductive system are:

   a) outside the body?

   b) inside the body?

testis one of a pair of organs that produce sperm

urethra a duct through which urine is expelled from the bladder and out of the body

penis the part of the male reproductive system through which sperm passes out

ovary one of a pair of organs in which ova (eggs) are produced

uterus the part of the female reproductive system, sometimes called the womb, in which a fertilised ovum becomes embedded and develops into an embryo

vagina the part of the female reproductive system that is open to the outside

Specialised animal cells

We are learning how to:

- recognise the relationships between specialised cells, tissues, organs and organ systems
- identify specialised animal cells.

Specialised animal cells

The various types of cells in the human body have different shapes and appearances. This is because they have different jobs to do.

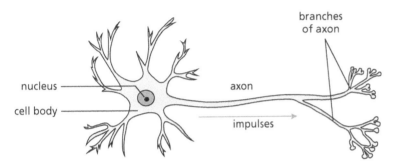

FIG 3.23.1 Nerve cell, or neuron

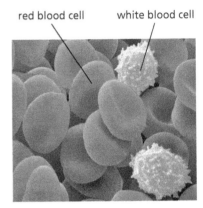

FIG 3.23.2 Red and white blood cells

Nerve cells, or **neurons**, carry electrical impulses from one place to another in the body.

Red blood cells carry oxygen around the body. They are red because they contain a pigment called haemoglobin, and have a characteristic 'doughnut' shape.

White blood cells attack and kill invading germs that would make us ill. One type of white blood cell kills germs by releasing chemicals, while another wraps itself around the germs and digests them.

Smooth muscle is found in many places in the body. It is different from the striated muscle found in the arms and legs. **Smooth muscle cells** are long and thin. The muscle is able to contract by making the cells shorter.

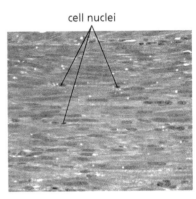

FIG 3.23.3 Smooth muscle cells

The **sperm** is the male reproductive cell. After intercourse, the sperm must swim up the female reproductive system until it finds an ovum. It does this by wriggling its tail, or flagellum.

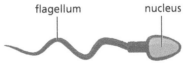

FIG 3.23.4 Male reproductive cell, or sperm

The **ovum** (plural ova) is the female reproductive cell. Ova cannot move like sperm, but are much larger. Each ovum contains yolk, which will nourish the zygote formed when a sperm penetrates the cell.

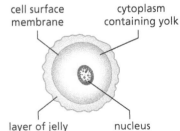

FIG 3.23.5 Female reproductive cell, or ovum

Activity 3.23.1

Animal cells, tissues, organs and systems

You will not need any equipment or materials for this activity. Here is what you should do:

1. Copy and complete Table 3.23.1 with the names of the missing cells, tissues, organs and systems.

Cells	Tissue	Organ	System
			Respiratory
	Heart muscle		
		Kidney	
Sperm, ovum			

TABLE 3.23.1

Activity 3.23.2

Job application sheet for specialised cells

Design a sheet for specialised cells applying for jobs in the body. List the cells and answer the following questions:

1. What do you look like? (Provide a drawing)
2. Which other cells/structures have you worked with?
3. In which department/organ would you prefer to work?
4. What are the special skills that you bring to the job?

Check your understanding

1. Fig 3.23.6 shows some of the cells in the lining of the throat.

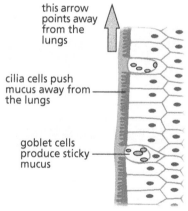

this arrow points away from the lungs

cilia cells push mucus away from the lungs

goblet cells produce sticky mucus

FIG 3.23.6

a) Name two types of specialised cell shown in Fig 3.23.6.

b) Suggest how these types of cell prevent particles that we breathe in from entering the lungs.

Key terms

nerve cell a cell that carries nerve impulses

neuron another name for a nerve cell

red blood cell a blood cell that carries oxygen around the body

white blood cell a blood cell that fights germs which invade the body

smooth muscle cells the type of cells from which heart tissue is formed

sperm the human male sex cell (plural 'sperm')

ovum the human female sex cell (plural 'ova')

Plant systems

We are learning how to:

- recognise the relationships between specialised cells, tissues, organs and organ systems
- identify systems in a flowering plant.

Plant systems >>>

Plants also have systems. The part that you can see above the ground is the shoot system, and the part you cannot see, because it is buried in the soil, is the root system. Each system has a different role.

All flowering plants have five important organs: **roots**, a **stem**, **leaves**, **flowers** and **fruits**. Each of these is important to the survival of the plant.

Roots

The roots of a plant are buried in the soil. Roots:

- hold the plant in the soil, so it is not blown about in the wind
- absorb water containing minerals from the soil
- can store starch for the plant.

Plant roots often have lateral root branches.

Roots have root hairs on their surface to absorb water.

Stem

The stem of a plant connects the roots to the leaves and flowers, above the ground. The stem:

- supports the parts of the plant above the ground
- allows substances to move around the plant.

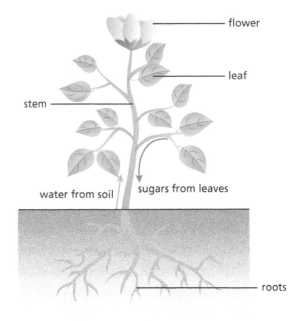

FIG 3.24.1 Parts of a flowering plant

Labels: flower, leaf, stem, water from soil, sugars from leaves, roots

Activity 3.24.1

Examining the tubes that carry water up a stem

Here is what you need:

- Solution of red ink (several drops of ink to water in a container)
- Stalk of celery, Balsam (lady slipper) or Impatiens.

Here is what you should do:

1. At the start of a lesson, take a freshly gathered celery stalk. Cut off the bottom and stand the stalk in red ink solution.

2. Near the end of the lesson, take the stalk out of the ink solution. Cut across the middle with a sharp knife.

3. Examine the cut part of the celery. Can you see any red dots where the stalk has drawn up the ink?

Leaves

The leaves are usually the most obvious part of a plant.

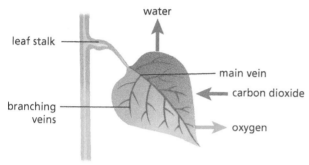

FIG 3.24.2 Parts of a leaf

Leaves are green because they contain chlorophyll, which is a green pigment.

The leaves:

- make nutrients by the process of photosynthesis

- allow the plant to absorb carbon dioxide for photosynthesis and release oxygen formed in this process

- lose water vapour in a process called transpiration.

Flowers

Flowers are the organs of sexual reproduction of a flowering plant. When a plant reproduces, it forms fruits and seeds.

Fruits

Fruits contain the seeds of the plant. They are sometimes eaten by animals, which helps the plant to disperse its seeds.

Check your understanding

1. Why is it better for a plant to have a network of many roots spread out in the surrounding soil than just a single root?

2. Why are the stem and leaves of a plant green, but the roots are not?

Fun fact

Not all groups of plants have flowers. For example, ferns do not have flowers.

FIG 3.24.3 Ferns in the Sangre Grande forest

Ferns reproduce in a different way to flowering plants. They produce spores instead of seeds.

Key terms

root the part of a plant below the ground that anchors the plant, and absorbs water and nutrients from the soil

stem the part of a plant above ground that has leaves, and sometimes produces flowers

leaf the flat green part of a plant that traps sunlight and carries out photosynthesis

flower the organ of sexual reproduction in plants

fruit the part of a plant that forms around the maturing seeds of the plant

Specialised plant cells

We are learning how to:

- recognise the relationships between specialised cells, tissues, organs and organ systems
- identify specialised plant cells.

Specialised plant cells »»

Plants have specialised cells in the same way as animals do. Each type of cell carries out a particular function within the plant.

Palisade cells

Palisade cells are found in leaves, beneath the outer epidermis, and often make up most of the leaf.

Palisade cells are vertically elongated and contain chloroplasts. The palisade cells absorb most of the light energy that the leaf uses in photosynthesis.

Guard cells

Guard cells are in the underside of leaves. A pair of **guard cells** surround each opening, or **stoma** (plural **stomata**).

The guard cells help to regulate the rate at which a plant loses water through its leaves, by opening and closing the stomata.

When the guard cells absorb water they swell and the stoma opens, allowing the plant to lose water. When the guard cells lose water they shrink. This closes the stoma, preventing the plant from losing any more water.

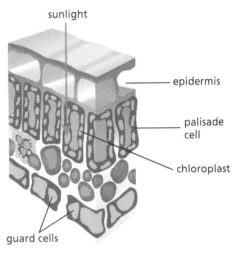

FIG 3.25.1 A section through the leaf of a plant, to show the specialised types of cell

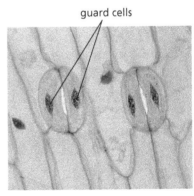

FIG 3.25.2 Guard cells in the underside of a leaf

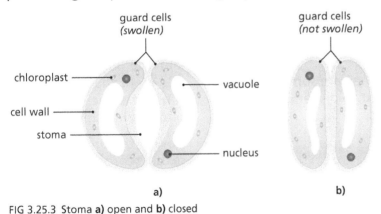

FIG 3.25.3 Stoma a) open and b) closed

Activity 3.25.1

Plant organs and their functions

You will not need any equipment or materials for this activity.

Here is what you should do:

1. Copy and complete Table 3.25.1 with the missing organs and functions.

Organ	Function
Leaves	
	Sexual reproduction
Stem	
	Absorbs water and nutrients
	Contains seeds

TABLE 3.25.1

Root hair cells

Root hair cells are present in the roots of plants. The root hairs are not hairs like you have on your head, but a hair-shaped part of a root cell that grows out into the soil.

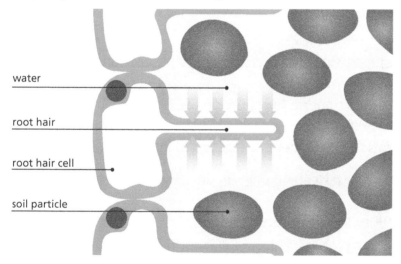

water

root hair

root hair cell

soil particle

FIG 3.25.4 Root hair cell

The root hair greatly increases the surface area of the root, making it easier for the plant to absorb water from the soil.

Check your understanding

1. With the help of suitable diagrams, explain how guard cells control the loss of water from a leaf.

Fun fact

Plant cells have a rigid cell wall made of cellulose that surrounds them. People cannot digest cellulose, so when we eat plant material, the cellulose remains. This undigested cellulose is the main constituent of fibre, or roughage, in our diet.

Key terms

palisade cell a cell in the leaf of a plant that contains chloroplasts

guard cell one of a pair of cells that surround a stoma, and open and close it

stoma an opening in a leaf that allow gases to pass in and out

stomata plural of stoma

root hair cell hair-shaped cells at the end of the roots, which have a large surface area to assist absorption

Review of Cells and organisms

- All organisms exhibit seven characteristics that we associate with living things. These characteristics are:
  - Growth
  - Respiration
  - Irritability
  - Movement
  - Nutrition
  - Excretion
  - Reproduction.

  If we take the first letter of each of these characteristics, it spells GRIMNER.

- All organisms are composed of cells.
  - Unicellular organisms consist of only a single cell.
  - Multicellular organisms consist of many cells.

- Cells are so small we can only see them with the aid of a microscope. In order to make the parts of a cell easier to see, we stain specimens of cells with dyes.

- A simple animal cell consists of:
  - a nucleus, which controls the processes that go on in the cell and contains chromosomes
  - cytoplasm, in which there are mitochondria that provide the cell with energy, glycogen grains and small vacuoles
  - a cell membrane, which controls the movement of substances into and out of the cell.

- A simple plant cell is similar in structure to an animal cell, but it contains some additional features:
  - A rigid cell wall made of cellulose that gives the cell shape
  - Chloroplasts containing the green pigment chlorophyll, which traps the energy needed for photosynthesis from sunlight
  - Starch grains in place of glycogen grains
  - Large vacuoles in place of small vacuoles.

- Both animal and plant cells have a cell membrane. A plant cell has an additional cell wall.

- Cells are the building blocks from which all organisms are formed. Complex organisms contain many different types of cell. These cells are arranged in tissues, the tissues form parts of an organ, and the organ forms part of an organ system.

  cells ⟶ tissues ⟶ organs ⟶ systems ⟶ organism

- The human body contains a number of important systems. Table 3.R.1 shows the main organs in some of these systems.

System	Main organ(s)
Circulatory system	Heart
Respiratory system	Lungs
Digestive system	Stomach, intestines, liver
Excretory system	Kidneys, skin, lungs
Skeletal and muscular system	Skeleton, muscles
Reproductive system	Testes, ovaries

TABLE 3.R.1 Systems and main organs of the human body

- Complex organisms have specialised cells that have particular functions within the organism.

- Specialised cells in people include: neurons, or nerve cells; red and white blood cells; sex cells (sperm and ovum); and smooth muscle cells.

- Specialised cells in a flowering plant include: palisade cells, guard cells, root hair cells, sex cells (pollen and ovum).

Review questions on Cells and organisms

1. Fig 3.RQ.1 shows some birds in a nest.

List the seven characteristics of birds that indicate to you that they are living organisms and not non-living objects. Write a sentence about each characteristic.

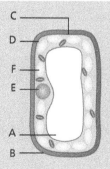

FIG 3.RQ.1

2. Fig 3.RQ.2 shows a typical plant cell.

a) Name parts A–F on the diagram.

b) Say whether each of the following statements about cells is true or false:

i) Animal cells have a cell wall made of cellulose.

ii) The nucleus controls many of the processes in the cell.

iii) Mitochondria are found in the cytoplasm.

iv) Plant cells store food as grains of glycogen.

FIG 3.RQ.2

3. Table 3.RQ.1 lists some cell parts. Copy and complete the table, as follows. Decide whether each part is found: only in animal cells (write A next to it in the table), only in plant cells (write P) or in both types of cell (write B).

Part of a cell	Write A, P or B
Cell membrane	
Cell wall	
Chloroplast	
Cytoplasm	
Nucleus	
Large vacuole	
Mitochondria	
Glycogen grains	

TABLE 3.RQ.1

4. Fig 3.RQ.3 shows a unicellular organism called *Euglena*.

Euglena lives in water and is able to move using a whip-like flagellum.

a) What does 'unicellular' mean?

b) Suggest one way in which *Euglena* is like:

  i) a plant

  ii) an animal.

c) Suggest why organisms like *Euglena* are no longer classified as plants or animals, but are placed in a separate group.

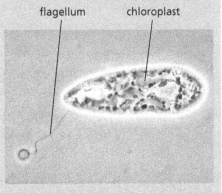
flagellum   chloroplast

FIG 3.RQ.3

5. A person's breathing rate is the number of times they breathe in and out each minute. Table 3.RQ.2 shows the range of breathing rates for people of different ages.

a) What general trend do the data show?

b) Present the data as a bar chart. Use the height of the bars to show the breathing rate.

Age in years	Typical breathing rate in breaths per minute
Up to 1	30–40
1 to 3	25–35
3 to 6	20–30
6 to 12	18–26
Over 12	12–20

TABLE 3.RQ.2

6. Fig 3.RQ.4 shows the pulse rates of a fit person and an unfit person before, during and after exercise.

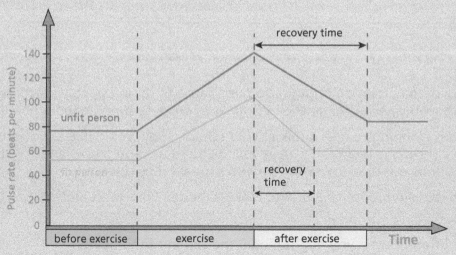

FIG 3.RQ.4

a) What was the pulse rate of the unfit person:

  i) before exercise?

  ii) at the end of the exercise period?

b) How does the pulse rate of a fit person compare with the pulse rate of an unfit person:

  i) at rest?

  ii) when exercising?

c) i) What difference is there between the recovery time of a fit person compared with that of an unfit person?

  ii) Suggest a reason for the difference in **i)** above.

7. a) Name the parts A–E of the female reproductive system shown in Fig 3.RQ.5

b) In which of parts A–E:

  i) are eggs released?

  ii) do fertilised eggs usually develop during pregnancy?

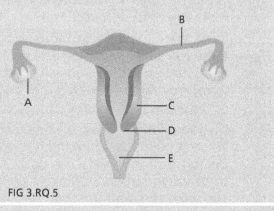

FIG 3.RQ.5

8. Fig 3.RQ.6 shows two specialised cells from the human body.

a) Identify cells A and B.

b) Describe one way in which the structure of each cell is suited to its role in the body.

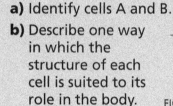

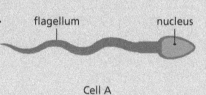

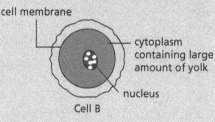

FIG 3.RQ.6

125

Making a model of the heart

The Heart Foundation of Jamaica is an organisation dedicated to raising people's awareness of heart disease and how they can change their lifestyles in order to become more healthy.

One of their speakers is planning to give a talk at your school. They would like a model to demonstrate the working of the heart. They have asked you to use your knowledge of the structure of the heart and your practical skills to make a model they can use.

1. You are going to work in groups of 3 or 4 to make a model. Your tasks are:

- To review how the heart works.
- To consider different ways you might model the action of the heart.
- To make a list of the materials and tools you will require and obtain them.
- To build a prototype of your model.
- To try out your model.
- To modify your model on the basis of the trial.
- To compile a report on how your model was constructed and how it works.

a) Look back through this unit and make sure you are familiar with the structure of the heart and understand how it works.

b) How might you construct a model of the heart?

The heart has four chambers. You could represent each chamber by a thick polythene bag. The arteries and veins that carry blood to and from the heart could be represented by coloured rope.

If you want to be a little more ambitious you can try to model the pumping action of the heart using plastic water bottles. You will need to devise some way of controlling the movement of liquid between the different parts of your heart.

Give some thought as to what sort of model you think you can build. Bear in mind it is going to be used for demonstration purposes as part of a talk so it needs to be easy to operate. Make a list of the materials you will need, and the tools you will need for your construction.

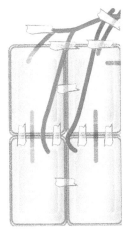

FIG 3.SIP.1 Polythene bags and string

c) Make your model and try it out. Better still, get someone else to try it out and ask them if they find it easy to operate. Take some pictures of the construction and the trial which you can use in your report.

Identify and problems in the construction and any difficulties in operation. Modify your model to overcome these problems and make the model as easy as possible to use.

d) Compile a report on how you built your model and demonstrate how it works. Use photographs and other evidence to explain why you modified your original construction and what improvements you were able to make as a result of the trial.

FIG 3.SIP.2 Plastic bottles, tubes and clips

Making a simple microscope

2. Materials: thin piece of wire, petroleum jelly, water, a page with writing on it.

a) Research the problem: use the internet to research designs for simple microscopes.

b) Design requirements: What are the characteristics that the microscope must have for you to be able to see through it?

c) Design constraints:

- Are there any limitations that will be caused by the resources that you have been given?

d) Brainstorm possible designs.

- Discuss in your groups possible designs for your microscope.
- Chose three possible designs that will meet your design requirements.
- Make sketches of these three designs.

e) Choose the design that you think will work best and give reasons for your choice.

f) Test and evaluate your solution

- Construct your microscope based on your design. Test to see if it meets your requirements.

g) Share your results with the other groups in the class.

h) Are there any improvements that you could make to your design?

i) Enjoy using your microscope.

Unit 4: Energy

We are learning how to:

• understand what energy is
• tell different forms of energy apart.

What is energy? »

Energy is the ability to do work.

Although all energy is the same, it has different effects in different contexts. Different forms of energy do different things.

For example, a television produces **light** energy, which we can see, and **sound** energy, which we can hear. It also produces some **heat** energy that we can feel if we touch the back of the device after it has been switched on for a short time. Light, sound and heat are all energy, but they are different because the energy does different things.

Our bodies have senses that allow us to detect some forms of energy. Our eyes detect light energy, our ears detect sound energy, and our skin detects heat energy.

We are only aware of other forms of energy because we can sense the effects they have. For example, we cannot see **electrical energy** in a wire, but we can sense the effect of electricity in a circuit. When an electric current passes through a bulb it lights up, which we can see, and it gives out heat, which we can feel.

Forms of energy include: heat energy, light energy, sound energy, electrical energy, **chemical energy**, **nuclear energy**, **potential energy** and **kinetic** (movement) **energy**.

FIG 4.1.1 A television produces light energy, sound energy and heat energy

Fun fact

The Sun produces twice as much energy in one hour as the entire population of the Earth uses in one year.

FIG 4.1.3 The Sun

Only a small fraction of the energy produced by the Sun reaches the Earth.

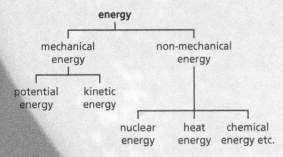

FIG 4.1.2 Forms of energy

Energy can be classified as mechanical or non-mechanical (see Fig 4.1.2).

Mechanical energy includes kinetic energy (movement energy) and potential energy (stored energy).

Non-mechanical energy includes light, heat, sound, electrical energy and nuclear energy.

Activity 4.1.1

Investigating devices associated with different forms of energy

Here is what you need:

- The photographs in Fig 4.1.4
- Your teacher may also provide some devices or pictures for you to examine.

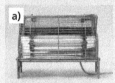 

FIG 4.1.4

Here is what you should do:

1. For each device, write the name of one form of energy that you associate with it.

2. Write your results in a table. On one side write the name of the device, and on the other write the name of the form of energy.

Check your understanding

1. What forms of energy are associated with the three pictures in Fig 4.1.5?

a)  b)  c)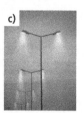

FIG 4.1.5 Different forms of energy

Key terms

light a form of energy that we can see with our eyes

sound a form of energy that we can hear with our ears

heat a form of energy that is transferred as a result of difference in temperature

electrical energy a form of energy easily conducted by metals and easily converted to heat, light and other forms

chemical energy a form of energy in food and fuel

nuclear energy a form of energy obtained from nuclear reactions

potential energy a form of energy stored in an object by virtue of its position, composition or shape

kinetic energy the energy an object has because it is moving

Potential energy

Potential energy >>>

Potential energy is one form of mechanical energy: it is energy that is stored in some way in a system. Potential energy may be classified as gravitational, chemical or elastic potential energy.

If an object is held in the air and released, it will fall to the ground without anyone exerting a force on it (see Fig 4.2.1). It is pulled towards the ground by the force of gravity. Any object above the ground is said to have **gravitational potential energy**.

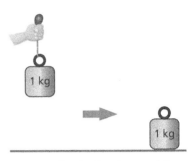

FIG 4.2.1 Gravitational potential energy

Food contains chemicals that are broken down in the digestive system. The products of digestion provide energy for the many processes in the body. Similarly, fuels contain chemicals that release energy when they are burned. Both food and fuels are examples of **chemical potential energy** (see Fig 4.2.2).

FIG 4.2.2 Chemical potential energy in: a) food b) fuel

If we place a weight on a spring, the spring gets longer (see Fig 4.2.3). The extended spring has **elastic potential energy**. If we remove the weight, the spring will return to its original shape.

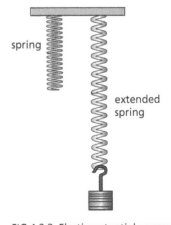

FIG 4.2.3 Elastic potential energy

Investigating elastic potential energy

Here is what you need:

- Elastic band
- Matchstick
- Clothes peg
- Measuring tape.

Here is what you should do:

1. Look at Fig 4.2.4. Stretch the elastic band over the length of the clothes peg.

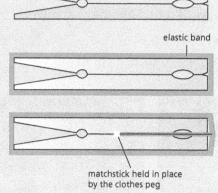

clothes peg

elastic band

2. Open the clothes peg and, making sure that the clothes peg is not pointing at anyone, push the matchstick into it. This pushes part of the elastic band back.

matchstick held in place by the clothes peg

3. Close the clothes peg. FIG 4.2.4

4. Make sure the clothes peg is not pointing at anyone. Open the clothes peg.

5. Measure how far the matchstick travels.

6. Repeat Steps 1–5 with some different elastic bands.

7. How will you be able to tell which elastic band stores the most potential energy?

Key terms

gravitational potential energy the energy an object has by virtue of its position above the ground

chemical potential energy the energy stored in a substance due to the chemicals it contains

elastic potential energy the energy stored when an elastic object, such as a spring, is deformed (stretched or compressed)

Check your understanding

1. Fig 4.2.6 shows a boulder on the edge of a cliff.

   a) What type of energy does the boulder have?

   b) Explain your answer.

FIG 4.2.6 Balanced Rock, Colorado, US

Kinetic energy

We are learning how to:
- describe how kinetic energy arises
- investigate how gravitational potential energy is converted to kinetic energy and back again.

Kinetic energy >>>

Kinetic energy is the energy associated with movement. Any moving object has kinetic energy.

Objects that move slowly, like snails, have only a small amount of kinetic energy.

FIG 4.3.1 **a)** Snail – small amount of kinetic energy. **b)** Dragster – large amount of kinetic energy

Objects that move quickly, like dragsters, have a large amount of kinetic energy.

A **pendulum** consists of a small weight attached to a string. The string is fixed at the opposite end to the weight, so the weight can swing to and fro.

When a pendulum swings, the energy of the pendulum bob continually changes between **gravitational potential energy** and kinetic energy.

Activity 4.3.1

Converting between potential energy and kinetic energy

Here is what you need:

- Small weight
- String
- Stand and clamp.

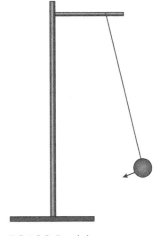

FIG 4.3.2 Pendulum

Here is what you should do:

1. Make a pendulum by tying a small weight to a length of string. Hang your pendulum on a stand and clamp, or in some other suitable place.

2. Hold the pendulum to one side so it is stationary (point A in Fig 4.3.3) and then release it.

3. Watch the motion of the pendulum as it swings to and fro through points A, B and C.

4. Copy and complete Table 4.3.1 with position A, B or C. Assume that the pendulum does not lose any energy as it swings.

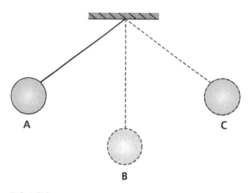

FIG 4.3.3

	Gravitational potential energy	Kinetic energy
Maximum		
Minimum		

TABLE 4.3.1

5. How do you know that the pendulum *does* lose energy as it swings?

At the start of the swing, the bob is stationary, so it has no kinetic energy. It is also at the maximum height above the ground or the table, so its gravitational potential energy is at a maximum.

When the bob is midway through its swing, the gravitational potential energy is at a minimum value. The lost potential energy has been converted to kinetic energy. The kinetic energy is at its maximum value.

As the bob continues through its swing, the gravitational potential energy starts to increase. This energy must come from somewhere: kinetic energy is being converted back to potential energy.

Check your understanding

1. Draw diagrams to show the position of a pendulum bob when:

   a) its kinetic energy is greatest

   b) its gravitational potential energy is greatest.

Fun fact

If there was no loss of energy, a pendulum bob would continue to swing for ever. In reality, a small amount of energy is lost during each swing, because of friction with the air and friction where the string is suspended. The height to which the bob swings slowly decreases until eventually the pendulum stops.

Key terms

kinetic energy the energy an object has because it is moving

pendulum a weight hung on a rod or string so that it can swing freely

gravitational potential energy the energy an object has by virtue of its position above the ground

Heat and light energy

We are learning how to:

- investigate heat and light energy
- investigate how light travels in straight lines.

Heat and light energy 〉〉〉

Heat and **light** are two forms of **energy** that we often experience together.

FIG 4.4.1 The Sun provides heat and light

The Sun is a major source of heat and light. Daytime is light and warm. Nighttime is dark and cooler.

A light bulb contains a thin wire called a filament. When electricity passes through the filament, the wire gets so hot that it gives out light.

FIG 4.4.2 A light bulb provides heat and light

FIG 4.4.3 Burning fuels produce heat and light

When we burn fuels like wood, both heat and light energy are produced.

Transforming light energy into electrical energy and back into light energy

Here is what you need:

- Solar cell
- Lamp
- Wires.

Here is what you should do:

solar cell

lamp

FIG 4.4.4

1. In an area of good light, for example near a window, make the circuit shown in Fig 4.4.4 A solar cell transforms light energy into electricity.

2. How does a lamp provide evidence that electricity flows around the circuit?

3. Cover the solar cell so it doesn't receive any light. Does the lamp light up?

4. Carry your circuit to different places in the laboratory where the light intensity varies from very dim to very bright and note the brightness of the lamp.

5. Comment on the link between the light intensity and the amount of energy generated by the solar cell.

6. What is the relationship between the light intensity and the brightness of the lamp?

Light and heat travel from place to place as waves. In Activity 4.4.1, you discovered that light only travels in straight lines. This is sometimes referred to as 'rectilinear propagation'.

Check your understanding

1. Copy and complete the following sentences:

    a) Heat and light are forms of ....

    b) Heat and light are both given out when a .... burns.

    c) Heat and light travel as ....

    d) Light can only travel in ....
    ....

Key terms

heat a form of energy that is transferred as a result of difference in temperature

light a form of energy that we can see with our eyes

energy the ability to do work

Sound energy

Sound energy »

Sound is a form of **energy** that we can detect with our ears. **Sound** is produced when an object vibrates. It passes through a medium as sound waves.

We are most used to hearing sounds that travel through air, but sound can also travel through liquids and even solids. If you put your ear to one end of your table and ask a classmate to scratch the other end, you will be able to hear the sound travelling through the table.

FIG 4.5.1 Sound travels through air

FIG 4.5.2 Sound travels through solids

Table 4.5.1 shows the speed of sound in different media. Sound travels fastest in solids and slowest in gases.

Medium	Example	Speed of sound in m/s
Gas	Air	340
Liquid	Seawater	1500
Solid	Steel	5000

TABLE 4.5.1

The space between the Earth and the Sun is a vacuum. It does not contain any solids, liquids or gases. Many massive violent explosions take place on the surface of the Sun each year. We cannot hear these explosions on Earth, because sound cannot travel through the vacuum of space.

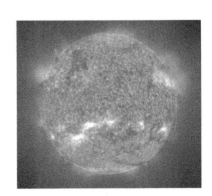

FIG 4.5.3 Sound cannot travel across a vacuum; the nuclear reactions taking place in the Sun are very loud, but we cannot hear them

Activity 4.5.1

Making a string telephone

Here is what you need:

- Empty can with the top removed × 2
- Length of string
- Hammer
- Small nail.

Here is what you should do:

1. Place an empty can on your desk, open end downwards.

2. Use a small nail and a hammer to make a small hole in the end of the can. The hole needs to be just large enough for string to pass through.

3. Repeat Step 2 with a second can.

4. Thread string through the hole in the first can and tie several knots in it so that the string cannot be pulled out.

5. Thread the other end of the string through the second can and tie knots to keep it in place.

6. With a partner, move the cans apart until the string is taut. One person should put the can to their mouth and say something while the other person puts the can to their ear to listen.

7. How do the sound waves travel from one can to the other?

Check your understanding

1. How does an object produce sound?

2. Through which type of material (solid, liquid or gas) does sound travel:

   a) fastest?

   b) slowest?

3. Why can people on Earth not hear loud explosions that occur on the Sun?

Fun fact

Sound waves can be used to locate objects under water.

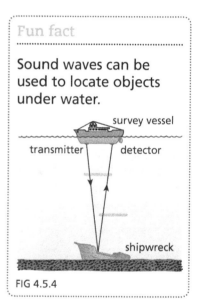

FIG 4.5.4

Key terms

energy the ability to do work

sound a form of energy that we can hear with our ears

Electrical energy

Electrical energy >>>

Electrical energy, or **electricity**, is a very convenient and useful form of energy.

FIG 4.6.1 Electricity power lines

Electricity is generated from other forms of energy in power stations. It is sent to homes and factories along cables, which are either suspended on poles or buried underground.

FIG 4.6.2 Electricity meter

Electricity is supplied to homes and businesses through electricity meters that record how much electrical energy has been used. Consumers must pay for the electrical energy they use at the end of each month.

Inside the home, electricity is supplied to each room by wires. Lights are turned on and off by switches, while electrical appliances are made to work by plugging them into sockets.

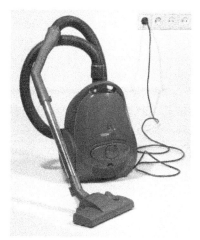

FIG 4.6.3 Electricity sockets

Most homes have a range of electrical appliances that change electrical energy into other forms of energy. You will learn more about these later in this section.

Activity 4.6.1

Survey of electrical appliances

Here is what you should do:

1. Do a survey of the electrical appliances in your home.

2. Place these devices into different groups according to what sort of energy an appliance releases: heat, electrical energy, sound, light or any other.

   Some devices might fit into more than one group. For example, a hairdryer releases heat but it also has a fan that releases kinetic energy.

Check your understanding

1. Where is electricity generated?

2. How is electricity sent from where it is generated to where it is needed?

3. How does the electricity company know how much electrical energy each consumer has used?

4. How is electricity supplied to an electrical appliance?

Fun fact

The Jamaican Public Services Company Limited (JPS) is responsible for the generation, transmission and distribution of electricity to consumers in Jamaica.

The JPS has a characteristic logo which people can easily recognise.

Jamaica Public Service Company Limited

CHANGING LIVES WITH OUR ENERGY

FIG 4.6.4 JPS logo

Key terms

electrical energy another term for electricity

electricity a form of energy that can be transferred from place to place along metal wires as an electric current

Nuclear energy

We are learning how to:

- produce nuclear energy using nuclear fusion or fission
- use the principle of conservation of mass and energy.

Nuclear energy »»

In order to account for the heat and light provided by the Sun, early scientists suggested the Sun might be a huge ball of burning coal. Such ideas were soon rejected for various reasons. Not until the discovery of **nuclear energy** were scientists able to explain the source of energy of the Sun.

The Sun's source of energy is a process called **nuclear fusion**. Under the conditions of high temperatures and a huge gravitational field, small atoms combine to form larger atoms.

FIG 4.7.1 The Sun produces massive amounts of energy

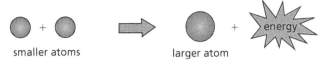

FIG 4.7.2 Nuclear fusion

The mass of the larger atom is slightly less than the mass of the two smaller atoms that formed it. The missing mass has been converted to energy.

FIG 4.7.3 A nuclear power station transforms nuclear energy into electricity

A different type of nuclear reaction called **nuclear fission** takes place in a nuclear power station. In this reaction, large atoms of a nuclear fuel like uranium break down to become smaller atoms.

larger atom   smaller atoms

FIG 4.7.4 Nuclear fission

The mass of the smaller atoms produced is slightly less than the mass of the large atom from which they were formed. The missing mass has been converted to energy.

Nuclear energy results from changes to the nuclei of atoms. During both nuclear fusion and nuclear fission, atoms of some elements are destroyed and atoms of other elements are created.

Nuclear energy is different to energy obtained from chemical reactions like burning. In chemical reactions, atoms are rearranged to make new substances but they are not destroyed or created.

Conservation of mass and energy

The principle of conservation of energy states that **energy cannot be created or destroyed**. Since nuclear reactions involve the loss of mass and the creation of energy, we need a new principle that takes nuclear reactions into account.

The principle must be modified to take into account both mass and energy. The principle of conservation of mass and energy states that:

> **The total mass and energy in a system is always the same.**

In other words, if some mass is lost, as in nuclear reactions, an equivalent amount of energy is released.

Check your understanding

1. What is the process by which energy is released in the Sun?

2. Briefly explain what happens during the process described in question 1.

3. How does the process described in question 1 differ from burning a fuel like wood?

Fun fact

Every second on the Sun, around 4 200 000 000 kilograms of mass is converted into energy. This sounds like a lot, but compared with the total mass of the Sun, which is 2×10^{30} kilograms, it actually is not very much. (2×10^{30} = 2 with 30 zeroes after it!) The Sun will continue to burn for billions of years to come.

Key terms

nuclear energy energy obtained when the nuclei of atoms undergo change

nuclear fusion the joining of two small nuclei to make one larger nucleus, with the loss of a small amount of mass

nuclear fission the division of one nucleus to form two smaller nuclei, with the loss of a small amount of mass

Non-renewable energy sources

We are learning how to:

• describe different non-renewable energy sources, including fossil fuels.

Non-renewable energy sources ⟩⟩

The sources of energy we use can be divided into two groups: renewable and non-renewable. If the energy source can be renewed by natural processes as fast as people use it, then it is **renewable**. If it cannot be renewed at this rate, then it is **non-renewable**.

Some sources of energy take many millions of years to form. People use the sources of energy up far more quickly than they can be replaced by nature. For this reason, these sources of energy are called non-renewable sources of energy.

Crude oil is not itself a useful source of energy but it is a raw material from which many fuels and other useful chemicals are made.

Natural gas is often found with crude oil. It is widely used for cooking and heating.

Coal is an important fuel in power stations for generating electricity. In Jamaica both fuel oil and natural gas are used to generate electricity.

FIG 4.8.1 The oil refinery at Kingston, Jamaica processes crude oil to create many useful products such as gasoline and other chemicals

FIG 4.8.2 Natural gas is a fossil fuel which is widely used for cooking

FIG 4.8.3 Coal

Crude oil, natural gas and coal are sometimes described as fossil fuels. They formed deep in the ground from the decomposition of plant and animal remains. The rocks that contain crude oil, natural gas and coal often have fossils of the organisms from which the fuels formed.

Fossil fuels are used to provide electricity. These types of power station produce large amounts of electricity but they have certain disadvantages.

FIG 4.8.4 An open-cast coal mine has a large impact on the environment

Some disadvantages of fossil fuels are:

- Removing a fossil fuel from the ground may damage the environment.

- Fossil fuels produce atmospheric pollutants when the fuels are burned.

- Only limited amounts of fossil fuels remain in the ground.

Check your understanding

1. Table 4.8.1 gives a rough estimate of how many years reserves of each fossil fuel will last. The values were obtained by dividing the known reserves of each fuel by the amount currently produced each year.

Fossil fuel	Number of years the fuel is expected to last
Coal	113
Crude oil	53
Natural gas	56

TABLE 4.8.1

   a) Reserves of which fossil fuel:
   i) are likely to run out first?
   ii) are likely to last for the longest time?
   b) Suggest one reason why these estimates may not be accurate.

Fun fact

The fuel for nuclear power stations is uranium.

FIG 4.8.5 Uranium fuel is contained in metal fuel rods

Uranium is also a non-renewable source of energy, but reserves of nuclear fuels will last much longer than reserves of fossil fuels.

Key terms

renewable renweable resources are natural such as wind, water and sunlight which are always available.

non-renewable non-renewable resources such as fossil fuels are not able to be restored.

Renewable energy sources

We are learning how to:

- describe different renewable energy sources.

Renewable energy sources 〉〉〉

Some sources of energy are continually replaced by natural processes at the same rate as they are used. They are called renewable sources of energy, and they will never run out.

Fresh water falls as rain and gathers in streams and rivers. Barriers called **dams** can be built across large rivers. The water is allowed to pass through tunnels in the dam wall, where it drives turbines to make electricity called **hydroelectricity**.

The Sun provides heat and light. **Solar panels** can turn solar energy into electricity.

In some parts of the world, **geothermal energy** is obtained from hot rocks near the surface. The energy converts water to steam, which can be used for heating and also to generate electricity.

On the coast, the water level rises and falls twice each day due to the movement of the tides. A tidal power station traps the water behind a barrier (a **tidal barrage**) at high tide and allows it to flow out at low tide. As the water flows out, electricity is generated from the **tidal energy**.

FIG 4.9.1 Dams hold back water

FIG 4.9.2 Solar panels gather energy from sunlight

FIG 4.9.3 Geothermal energy comes from the ground

FIG 4.9.4 A tidal barrage traps water at high tide

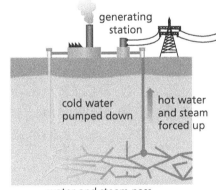

generating station

cold water pumped down

hot water and steam forced up

water and steam pass through cracks in the rocks

FIG 4.9.5 Cold water is heated as it passes through the hot rocks

Wind is the flow of air from one place to another. Wind has movement energy that can be used to do work. **Wind turbines** use energy from the wind to generate electricity.

Renewable sources of energy have both advantages and disadvantages.

Some advantages of renewable energy sources are:

FIG 4.9.6 Wind drives wind turbines, which generate electricity

- The energy is free.
- The energy will never run out.
- Making use of renewable energy causes relatively little environmental damage.
- Some devices are cheap to buy and install.
- Some devices are better suited than a power station to satisfying the needs of a small community.

Some disadvantages are:

- Some sources are not available all the time; for example, wind turbines only generate electricity when there is wind to drive them.
- Some sources provide relatively little energy; for example, it takes a large number of wind turbines to produce the same amount of electricity as a fossil-fuel-powered power station.
- Some sources are not suitable for all countries; for example, hydroelectricity is no use in a country where there are no fast-flowing rivers.

Solar water heaters use heat from the Sun, which is a renewable source of energy.

FIG 4.9.7 Solar water heater

Key terms

dam a wall built across a river to control the flow of water

hydroelectricity electricity that is generated using energy from moving water

solar panel a device that converts sunlight into electricity

geothermal energy heat energy that is obtained from the ground

tidal barrage a barrier across a tidal river to trap the water that flows in as the tide rises

tidal energy energy associated with the rise and fall of the sea level due to tides

wind turbine a device that converts kinetic energy from the wind into electricity

Check your understanding

1. From the examples of renewable energy sources given above, select the following, giving reasons for your choice:

   a) One source that would be suitable in Jamaica.

   b) One source that would not be suitable in Jamaica.

145

Biofuels

We are learning how to:

- describe different types of biofuel
- make a biogas generator.

Biofuels ⟩⟩⟩

Biofuels are the result of energy stored by photosynthesis – either directly, like **wood** obtained from trees, or indirectly, by animals eating plants and producing dung. They are an important source of energy in some countries.

In Africa, many people rely on wood and on **charcoal**, which is made from wood, as fuels to cook their food. Wood can be a renewable source of energy if new trees are planted to replace those that are used for fuel.

Biogas is an impure form of methane, made by fermenting animal dung. Biogas units can be small and serve one family, or they can be large and serve a small village. What remains of the dung after gas production can be used as a fertiliser in the soil to help crops grow and give better yields.

FIG 4.10.1 Wood is an important fuel in many parts of Africa

FIG 4.10.2 Biogas is a renewable source of energy

Activity 4.10.1

Making a biogas generator

Here is what you need:

- Plastic drink bottles (2 litre or larger) × 2
- Dung (goat or cow but not chicken)
- Strong, clear plastic bag
- Stopper × 3 (two 2-hole stoppers and one 1-hole stopper to fit the bottles and bag)
- Plastic tubing
- Weak solution of sodium hydroxide
- Small clamp like a spring clothes peg.

Here is what you should do:

1. Arrange the bottles and bag as shown in Fig 4.10.3. You might need a stand and clamp to support the bag.

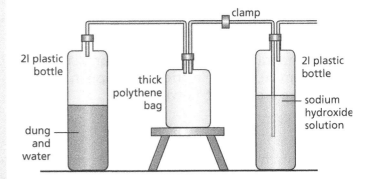
FIG 4.10.3

labels: clamp; 2l plastic bottle; thick polythene bag; 2l plastic bottle; dung and water; sodium hydroxide solution

2. Half fill the left-hand bottle with a mixture of dung and water. Half fill the right-hand bottle with sodium hydroxide solution.

3. Place stoppers in the bottles and polythene bag, and connect them with rubber tubing.

4. Place a clamp on the tube leading from the polythene bag to the bottle containing sodium hydroxide solution.

5. Put the apparatus somewhere warm and leave it for several days.

6. When the bag starts to swell up, remove the clamp and push bubbles of biogas through the sodium hydroxide solution.

7. With the help of your teacher, try to light the gas.

Notice that the biogas generator initially produces carbon dioxide, which is not flammable. When the biogas is bubbled through sodium hydroxide, the carbon dioxide should be removed. The sodium hydroxide container also prevents any flashback once the gas is lit. If your gas does not light first time, let the gas build up for another day and then try again.

Ethanol is a biofuel made by fermenting sugar or other plant material.

In some countries, ethanol is mixed with gasoline or used as an alternative fuel to gasoline for road vehicles (see Fig 4.10.4).

FIG 4.10.4 Ethanol can be used as an alternative to gasoline

Fun fact

Biodiesel is a renewable alternative to the diesel fuel obtained from crude oil. Biodiesel is made from vegetable and animal oils, and is therefore a biofuel.

FIG 4.10.5

Check your understanding

1. Charcoal is an important fuel in the Caribbean, where it is used to cook food.

FIG 4.10.6 Charcoal

   a) Carry out some research to find out how charcoal is obtained.

   b) Explain why charcoal is a biofuel.

Key terms

biofuel a fuel obtained from living matter

wood a fuel obtained from trees

charcoal a fuel obtained by heating wood in the absence of air

biogas a fuel gas made by the fermentation of animal waste

ethanol a chemical made by the fermentation of sugar that can be used as a fuel

Transforming energy

We are learning how to:

- investigate the transformation of energy from one form to another.

Transforming energy from one form into others ⟩⟩⟩

There are many examples in everyday life where one form of energy is changed into others. These changes are called **energy transformations** and they take place when work is done.

When fuels burn, the chemical potential energy they contain is converted into heat energy and light energy.

FIG 4.11.1 Energy transformations take place in our bodies

FIG 4.11.2 When fuels burn they produce heat and light energy

We can show this as a flow diagram:

chemical energy ⟶ heat energy + light energy

The food that we eat contains chemical potential energy. It is broken down to provide heat energy to maintain body temperature, and for other needs such as movement:

chemical energy ⟶ heat energy + kinetic energy

In an electric cell, chemicals produce electricity:

chemical energy ⟶ electrical energy

Many electrical appliances transform electrical energy into other forms of energy.

An electric iron transforms electrical energy into heat energy:

electrical energy ⟶ heat energy

FIG 4.11.3 An electric cell

FIG 4.11.4 An electric iron

Often, the transformation of electrical energy involves the production of more than one form of energy. For example, a television produces light and sound. It also produces a small amount of heat, which is lost to the surrounding air:

$$\text{electrical energy} \longrightarrow \text{light energy} + \text{sound energy} + \text{heat energy}$$

FIG 4.11.5 Television set

Activity 4.11.1

Investigating energy transformations

Here is what you need:

- Devices or pictures of devices to examine (see Fig 4.11.6).

a)  b)   c)

d)  e)

FIG 4.11.6

Here is what you should do:

1. For each device, decide how it changes energy.

2. Draw a flow diagram to show the energy transformation.

3. Make a classroom display to illustrate energy transformations.

Check your understanding

1. Draw a flow diagram to show the energy transformations that take place in each of the following:

   a) An electric kettle
   b) A bicycle bell
   c) A burning candle.

Fun fact

Many electrical appliances produce small amounts of unwanted heat and also, sometimes, sound energy. For example, the main energy transformation in a coffee grinder is from electrical energy to kinetic energy. But a coffee grinder also produces sound and heat, which is why it feels warm after someone has used it.

Key term

energy transformation a process in which one form of energy is changed into other forms of energy

149

Sankey diagrams

Sankey diagrams ⟩⟩

A Sankey diagram is a way of showing the proportions of different forms of **energy** released during an **energy transformation**. The proportion can be shown as an actual number of joules (J) or simply as a percentage of the starting energy.

Remember that energy can neither be created nor destroyed. When drawing a diagram to show an energy transformation, the amount of energy input will be exactly the same as the sum of the forms of energy output.

For example, in a filament light bulb, 90% of the electrical energy is transformed into heat energy and 10% is transformed into light energy. Filament bulbs are not very efficient at producing light. Most of the electrical energy is transformed into heat, which is why light bulbs get hot.

Fig 4.12.1 shows how the energy transformations in a filament light bulb are represented in a **Sankey diagram**. The width of the arrows is in proportion to the heat energy and light energy produced.

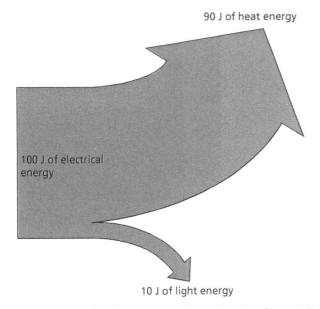

90 J of heat energy

100 J of electrical energy

10 J of light energy

FIG 4.12.1 Sankey diagram for the energy transformations in a filament light bulb

Activity 4.12.1

Drawing a Sankey diagram

Here is what you need:

- Squared paper
- Ruler.

Here is what you should do:

1. Read the following information about a hairdryer:

   When a hairdryer is operated, 10% of the electrical energy is transformed to kinetic energy to drive the fan, 40% is converted into sound and 50% into heat.

2. Start with a line on the left that is 10 squares high to represent 100% of the energy.

3. Draw a Sankey diagram to represent the information about the hairdryer.

We can also draw Sankey diagrams for energy transformations that release more than two forms of energy.

Fig 4.12.2 shows how the energy obtained by burning a fuel in a power station is transformed into other forms of energy. From this diagram we can see that less than half the energy in the fuel is transformed into electricity.

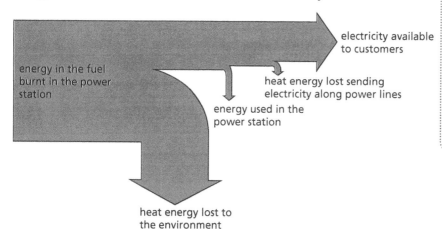

FIG 4.12.2 Sankey diagram for the energy transformations in a fossil-fuelled power station

Check your understanding

1. Use the Sankey diagram in Fig 4.12.2 to estimate what percentage of the energy contained in the fuel burnt in the power station is transformed in the different ways shown.

Key terms

energy the ability to do work

energy transformation a process in which one form of energy is changed into other forms of energy

Sankey diagram a diagram used to show energy transformations

Conservation of energy (1)

Conservation of energy – the Government ⟩⟩

The population of the world is growing every day, and people are continually looking for ways to raise their standard of living. This creates an ever-increasing demand for **energy**.

Energy conservation may be considered at the level of countries and also at the level of individuals. In considering energy conservation, you are really looking at ways of making more use of renewable sources of energy so that those energy sources that are not renewable can be conserved.

FIG 4.13.1 World demand for energy increases every year

Jamaica is heavily dependent on fossil fuels to generate electricity. Over 90% of the country's energy requirements are met by burning fossil fuels. This is expensive for the country, and it cannot go on for ever. Eventually reserves of fossil fuels will run out.

In 2009, Jamaica adopted a National Energy Policy which aims to provide 30% of the country's energy needs from renewable sources by 2030. At the same time, there is also a drive to use energy more efficiently to reduce demand.

At the moment, around 10% of the country's energy requirement comes from renewable energy sources including hydro, solar and wind turbine power plants. More energy will come from these sources in the future as different projects centred on renewable energy sources come to fruition.

FIG 4.13.2 Content Solar Power Plant in Clarendon

FIG 4.13.3 A wind farm like the one being built at Malvern in St Elizabeth

In 2016, the Content **Solar Power Plant** started operations. It is estimated that it will eventually provide 20 MW of power, which is enough to satisfy the needs of over 20 000 households. Each solar panel in the array uses sunlight to generate electricity.

A **wind farm** is being built at Malvern St Elizabeth. This area is hilly so the wind turbines can be sited where they get most wind. It is estimated that the wind farm will provide 36 MW of electrical power.

At present a feasibility study is being carried out to determine whether some of the country's rivers would be suitable for **hydropower plants**. The rivers being considered are: Morgan's River, Negro River, Spanish River, Rio Cobre and Martha's Brea.

FIG 4.13.4 A hydropower plant

Activity 4.13.1

Development of renewable energy sources in Jamaica

You will not need any equipment or materials for this activity. You should work in groups of 3 or 4.

Here is what you should do:

1. Carry out research to find out if any other sources of renewable energy are being developed in Jamaica. For example, are there any plans to harness energy from waves or moving water as a result of the tides?

2. Consider the advantages, and the disadvantages, of the different renewable sources described above. For example:

   a) Arrays of solar panels require land that could be farmed.

   b) Wind farms may detract from the natural beauty of an area.

   c) Building dams across rivers alters the local ecology.

3. Take part in a discussion on how Jamaica's renewable energy initiative should develop in the future. Include discussion of household use of alternate energy sources.

Check your understanding

1. Why is the wind a renewable energy source?

2. What energy transformations take place in a wind turbine?

3. Which of the five rivers mentioned in connection with hydropower is nearest to where you live?

4. Suggest some effects on the local environment of building a hydropower plant.

> **Fun fact**
>
> Not everyone agrees that wind farms are a good thing. Wind farms harness a renewable source of energy and do not pollute the atmosphere, but some people think they are eyesores that ruin the natural beauty of the landscape.

Key terms

energy the ability to do work

solar power plant device that converts energy from the Sun into electricity

wind farm collection of wind turbines that convert kinetic energy from the wind into electricity

hydropower plant device that converts kinetic energy from moving water into electricity

Conservation of energy (2)

We are learning how to:

- conserve energy ourselves by making better choices.

Conservation of energy – the individual >>

Governments decide a country's energy policy, but the individual also has a role to play in conserving energy.

Petrol and diesel are obtained from crude oil, which is a non-renewable source of energy. Every time we drive somewhere we use more of that energy source and also add to atmospheric pollution. There are times when people need to use cars and lorries, but there are also times when people can save energy by walking or cycling.

FIG 4.14.1 Cycling conserves energy and does not pollute the environment

Traditional light bulbs contain a metal filament that glows white hot to produce light. Light bulbs are very inefficient. Around 90% of the electricity they consume is wasted as heat energy.

Modern energy-saving bulbs work in a different way. Energy-saving light bulbs are often more expensive than traditional light bulbs, but they are still better value. They last longer and give out the same amount of light, but use less electricity.

Energy can be **conserved** by changing to energy-saving bulbs and also by remembering to switch lights off when you leave a room.

FIG 4.14.2 Modern bulbs use less electrical energy than filament lamps

Activity 4.14.1

How can I conserve energy?

Work in groups of four.

Here is what you should do:

1. In your group, discuss how you might be able to conserve energy. Make a list of your ideas.

2. Nominate one person in the group to be the spokesperson and share your group's ideas with the rest of the class.

3. Discuss your class' findings with the other members of your family. Challenge them to carry out these energy-saving ideas for one billing month.

4. Compare the energy used with that used in the previous month.

Much of the energy that countries consume is used when manufacturing materials. If people used less materials, there would be a significant saving in energy. Here are three things you can do – the 'three Rs':

- **Reduce** the quantity of manufactured goods you use. For example, do you really need a bag every time you buy something at the store?

- **Reuse** some items instead of buying new ones. For example, if you need a bag to carry things home from the shop, take the bag you were given last week. You do not need a new one every time.

FIG 4.14.3 Examples of materials that can be recycled

- **Recycle** materials so they can be made into new objects. If you have empty glass or plastic containers, or metal cans, those materials can be reused. Do not just throw them away.

Check your understanding

1. Table 4.14.1 gives some information about an old type of electric light bulb and a modern energy-saving bulb.

Type of bulb	Cost to buy	Cost to light the bulb for 10 hours each day	Number of days before needs to be replaced
Old	$3	$1	100 days
Modern energy-saving	$15	$0.40	300 days

TABLE 4.14.1

   a) How much more expensive is it to buy a modern bulb than an old-style bulb?

   b) For how many days would a person have to use a modern bulb in order to save the difference in the cost between an old bulb and a modern bulb?

   c) How much would a person save lighting a room for 300 days using modern bulbs rather than old bulbs?

Key terms

conserve to make something last longer by using less of it

reduce to make smaller

reuse to use an object again

recycle to use the material from an object to make a new object

Energy and the Caribbean countries

We are learning how to:

• conserve energy ourselves by making better choices.

Sources of energy now ▶▶▶

Like many other countries of the world, the countries in the Caribbean region are heavily dependent on fossil fuels. There are various reasons why individual countries, like Jamaica, have stated they will be trying to replace fossil fuels with renewable sources of energy over the coming decades. These include:

• Fossil fuels are expensive to import.

• Fossil fuels will eventually run out.

• Burning fossil fuels creates pollution.

FIG 4.15.1 Fossil-fuelled power stations pollute the atmosphere

Sources of energy in the future ▶▶▶

In lesson 4.13 you found out that Jamaica is planning to utilise different renewable sources of energy including wind, running water and sunlight. Other Caribbean countries are also taking steps to develop renewable energy. In this lesson we will look at some examples.

Trinidad and Tobago are carrying out a program to identify locations on the east coast of Trinidad where wind farms could be sensibly built. The west coast is preferred to the east because it receives strong winds from the Atlantic Ocean.

A number of Caribbean countries are investigating the possibility of harnessing **geothermal energy** from hot rocks deep under the ground.

FIG 4.15.2 Possible locations for a wind farm on Trinidad

The Bouillante Geothermal Power Station in Guadeloupe has been operating for about 30 years. At present, it has a power output of 15 MW but it is hoped to develop the power station to produce 45 MW by 2021.

Geothermal power requires the right sort of **geology** beneath the ground. At present Dominica and Nevis look like the most likely countries to benefit from this source.

One resource that the Caribbean is not short of is sunlight. In the previous lesson you learned about the Content Solar Power Plant in Clarendon; however, utilising solar power is not limited to Jamaica.

FIG 4.15.3 Bouillante Geothermal Power Station in Guadeloupe

FIG 4.15.4 Oriana Solar Farm, Puerto Rico

In 2016 the largest solar power station in the Caribbean was energised in Puerto Rico. It uses free energy from sunlight to produce 45 MW of electrical power. Many other Caribbean countries are looking at solar power as a means of reducing their dependence on fossil fuels in the future.

> **Fun fact**
>
> The Middle East has about half of the world's proven crude oil and natural gas reserves but countries in this region are keen to develop renewable energy sources like solar energy and wind power.

Activity 4.15.1

The energy debate in the Caribbean

You should work in groups of 3 or 4.

Take part in a class debate on the moral and social issues around energy use and energy sources in the Caribbean. Here are some proposals that can be discussed.

- If individuals used less energy there would be less demand.

- Renewable sources of energy are not capable of producing sufficient energy to replace non-renewable sources.

- There should be a Pan-Caribbean approach to energy generation and distribution rather than different countries doing their own thing.

Key terms

geothermal energy obtained from hot rocks deep underground

geology study of the Earth and its structure

Check your understanding

The bar chart shows the capacity for generating electricity using different energy sources in six Caribbean countries in 2015.

1. Which country is entirely dependent on fossil fuels to generate electricity?

2. Which country makes best use of renewable energy sources to generate electricity?

3. Approximately what percentage of Suriname's electricity is obtained from fossil fuels?

4. Predict how a similar bar chart drawn up in 2020 might differ from the one above.

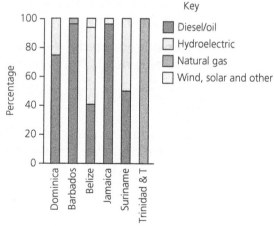

Capacity for generating electricity using different energy sources

Source: based on World Bank, IDB and IMF calculations

FIG 4.15.5

Review of Energy

- Energy is the ability to do work.

- Energy in different contexts is described as different forms of energy.

- Forms of energy include: heat, light, sound, electrical energy, chemical energy, nuclear energy, potential energy and kinetic energy.

- Potential energy is energy that is stored in some way and includes: gravitational potential energy, chemical potential energy and elastic potential energy.

- Kinetic energy is the energy that objects have when they move.

- In a swinging pendulum, energy is converted between gravitational potential energy and kinetic energy.

- Nuclear energy is the result of changes to the structure of atoms. It is the source of energy in the Sun and in nuclear power stations.

- Non-renewable energy sources are those that are not replaced by nature at the same rate as they are used up. They include coal, crude oil and natural gas.

- Renewable energy sources are those that are replaced by nature at the same rate as they are used up. They include hydroelectric energy (flowing water), solar energy, geothermal energy, tidal energy and wind energy.

- Biofuels are the result of energy stored by photosynthesis either directly, for example in wood obtained from trees, or indirectly, for example by animals eating plants and producing dung. Biofuels include biogas, wood, charcoal and ethanol.

- Energy can be transformed from one form into others. When this happens, work is done.

- Energy transformations can be shown as Sankey diagrams.

- The world is moving towards a greater use of renewable sources of energy as reserves of non-renewable energy sources are used up.

- Jamaica is currently developing renewable energy resources including solar power, wind power and hydroelectric power.

- In 2009 Jamaica adopted a National Energy Policy which aims to provide 30% of the country's energy needs from renewable sources by 2030.

- Individual people can make a significant contribution to reducing a country's demand for energy.

- All of the Caribbean countries are looking for ways of reducing their reliance on fossil fuels by developing the renewable energy sources which are available to them.

Review questions on Energy

1. **a)** What is the difference between potential energy and kinetic energy?
   b) Draw a diagram of a pendulum to show the position of the bob when:
      i) the potential energy is maximum
      ii) the kinetic energy is maximum.

2. State one form of energy that can be sensed by the:
   a) eyes   **b)** ears   **c)** skin.

3. **a)** Name the process by which energy is released in a nuclear power station.
   b) Explain how the release of energy from a nuclear fuel like uranium is different than from a chemical fuel like wood.

4. Table 4.RQ.1 shows some of the world's energy sources. They are arranged in order of the total amount of energy they provide.

Energy source	Relative amount of energy provided
Oil	Most energy supplied
Natural gas	
Coal	
Nuclear	
Hydroelectric	
Wind farms	Least energy supplied

TABLE 4.RQ.1

  a) From this table give:
    i) a non-renewable source of energy
    ii) a renewable source of energy.
  b) What is meant by a 'non-renewable energy source'?
  c) Name one other large-scale source of energy not given in the table.
  d) If a similar table was drawn up 100 years from now, suggest one way in which it would be different.

5. The following pie chart shows the proportions of energy a country gets from different sources.

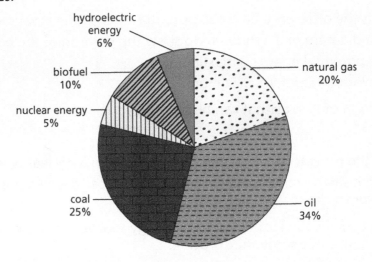

hydroelectric energy 6%

biofuel 10%

nuclear energy 5%

natural gas 20%

coal 25%

oil 34%

FIG 4.RQ.1

    a) Which source provides:

      i) most energy?

      ii) least energy?

    b) Explain why hydroelectric energy can be described as a renewable source of energy.

    c) Explain how the energy in biofuels originally came from the Sun.

6. The following energy transformations take place in different devices. For each transformation, name one example of a suitable device.

    a) electrical energy ⟶ heat energy

    b) chemical energy ⟶ heat energy

    c) electrical energy ⟶ kinetic energy

    d) kinetic energy ⟶ electrical energy

    e) electrical energy ⟶ light energy

    f) electrical energy ⟶ sound energy

7. A dairy farmer has decided to conserve energy by using a renewable energy source. He has three options:

    a) Biogas – he has cattle dung from his dairy cows.

    b) Hydroelectricity – a small river flows through his land.

    c) Wind turbine – there is a hill on his land where a wind turbine could be built.

Briefly explain how each of these could provide energy. Give one disadvantage of each, apart from the cost of starting up.

8. A wind turbine is to be built at Windy Cove. The following graphs show details about the conditions and the electricity that would be generated.

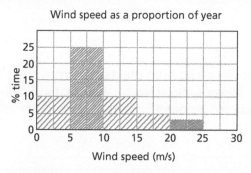

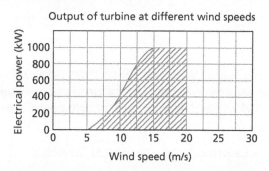

FIG 4.RQ.2

a) What is the maximum number of kilowatts of electricity the wind turbine will be able to generate?

b) Suggest a reason why:

   i) No electricity would be generated when the wind speed falls below 5 metres per second (m/s).

   ii) The wind turbine would be prevented from rotating when the wind speed rises above 20 m/s.

c) Predict for how many days the wind turbine would generate electricity.

Renewable sources for school electricity

The Principal wants your school to start producing some of its own electricity from suitable renewable energy sources. The target is to produce 30% of the school's electricity within 5 years. The Principal needs technical advice on the type of device the school should adopt to achieve this.

Your task is to provide this technical advice.

1. You are going to work in groups of 3 or 4. You are part of a school committee set up to consider this issue and make recommendations. These are the tasks:

- Find out how much electricity the school currently uses each year.
- Identify devices that could provide the school with electricity.
- Identify sites within the school grounds where devices could be located.
- Build a prototype of the device, which can be located at the designated location at the school and used to test the viability of your recommendation.
- Write a PowerPoint presentation in which you recommend which device the school should adopt and explain the reasons for your choice.

a) When carrying out your research and making your recommendations you need to be aware of the constraints and limitations associated with each type of device. The device that produces the most electricity might not be the one that is most suitable for the school to use. What factors should you consider in your research and recommendations? Discuss the possible factors in your group and make a list.

b) Research renewable sources of energy that can be used to produce electricity. Use the information in lesson 4.9 of your textbook to guide you. You might also find useful information by looking up other phrases like the 'solar furnace' and 'wave energy' on the Internet. What renewable sources of energy are being used or will soon be used in schools and colleges in different parts of Jamaica?

c) Consider the suitability of the different devices available for use within your school. Be aware of the constraints and limitations associated with each type of device. Consider issues such as:

- How much electricity is the school hoping to produce from renewable energy sources? That is, how much electricity does the school use each year and what is 30% of this amount?
- Which devices will be able to supply this amount of electricity?
- Is the location and the fabric of the school appropriate to a particular device? For example, is the school in a sheltered spot that receives little wind? Do the school buildings have roofs where solar panels could be positioned?
- Which devices are practical options when considering the cost of buying them and their maintenance?

Decide which renewable energy source you are going to recommend. Prepare reasons for your choice.

FIG 4.SIP.1 **a)** Solar furnace at Odeillo, France **b)** wave energy transformer floats **c)** wind turbines promotion in Jamaica **d)** solar panels at a classroom in St John, US Virgin Islands

d) Build a model of a device that uses the renewable source of energy you have chosen. For example, you might build:

- A solar panel connected in a circuit with a light-emitting diode (LED) that lights up when the sun shines on the solar panel.
- An electric motor with a propeller attached to the spindle. When the propeller is driven around by the wind, the motor becomes a generator and produces an electric current that can be detected using an ammeter.

Since both wind and sunlight are not available every hour of the day, you should think about how electricity can be stored by a capacitor on your prototype, and by a battery on the actual device.

e) Test your model to determine if it can carry out its designed function.

- Place your device in the location you have decided is suitable for it and monitor the amount of electricity being produced over a week.
- Compare this with the amount of electricity the school hopes to produce and decide how your prototype might be scaled up to meet this amount.
- Consider ways in which you might improve the efficiency of your prototype.

f) Give a PowerPoint presentation on what you have found and what you will recommend to the class. You should be prepared to explain why you are recommending this renewable energy source. Details of your trials with your device should be used.

Discuss possible improvements with the class and refine your recommendation and the design of your prototype as necessary.

Unit 5: Plant reproduction

We are learning how to:

• describe reproduction in plants.

Plant reproduction »

The growth cycle of a flowering plant

In the same way that animals are born, grow, reproduce and die, the lives of flowering plants also follow a cycle. The flower is the plant's organ for **sexual reproduction**.

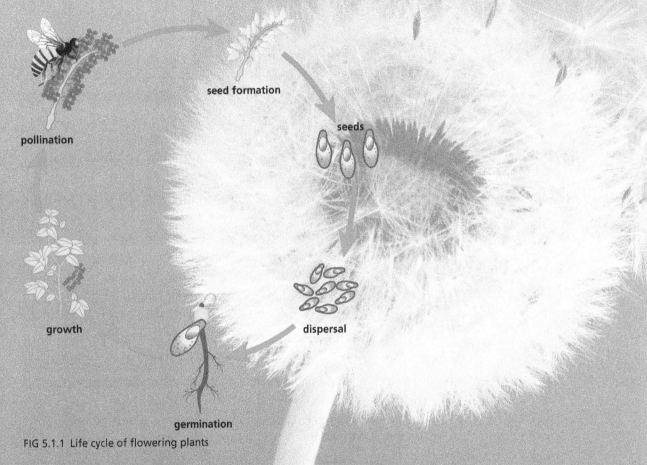

seed formation

seeds

pollination

growth

dispersal

germination

FIG 5.1.1 Life cycle of flowering plants

Each of the processes shown in the diagram is important if a plant species is to survive and prosper.

A seed contains sufficient food stores for a plant to begin to grow. During **germination**, the seed splits open and roots grow down into the ground while the stem grows upwards towards the light.

Before a plant can itself produce seeds, it must grow to a mature size. Some plants can germinate, grow and produce

FIG 5.1.2 The flower is the organ of sexual reproduction

seeds within a few months, while, for slow-growing plants, this process might take years.

The **flower** is the plant's organ of sexual reproduction. It contains both male and female parts, and male and female sex cells. The grains of pollen produced by a flower are its male sex cells. During **pollination** pollen is transferred to the female part of the same or a different flower. The transfer of pollen can come about in different ways.

Once pollen grains arrive on the female part of a flower they combine with the ovules, which are the female plant sex cells, and **fertilisation** takes place. Each pollen grain combines with an ovule to form a zygote, which will eventually become a seed.

A single flowering plant may have several flowers, and each flower may be capable of producing many seeds. If the seeds simply dropped and germinated in the ground around the parent plant there would soon be too much competition for essential factors like sufficient light, and nothing would flourish.

To prevent this, plants have developed different mechanisms for **dispersing** seeds well away from the parent plant, which increases their chances of survival.

In addition to sexual reproduction, many plants are able to produce offspring by **asexual or vegetative reproduction**.

Key terms

sexual reproduction involves male and female sex cells which may be provided by two parents

germination start of growth of a seed

flower organ of sexual reproduction in flowering plants

pollination transfer of pollen from the male parts of a flower to the female parts of the same or a different flower

fertilisation coming together of a male sex cell and a female sex cell to form a gamete

dispersal movement of fruits and seeds away from the parent plant

asexual or vegetative reproduction only involves a single parent

Flower structure

We are learning how to:

- identify the different parts of a flower
- identify the male and female parts of a flower.

Flower structure ▶▶▶

A flower has both male and female parts. The male part is called the **stamen** and the female part is called the **pistil**.

The stamen and pistil are at the centre of the flower where they are surrounded by the petals and above the nectaries, which provide sweet nectar for visiting animals. You will learn more about the importance of petals and nectaries in the next lesson.

Stamen and pistil

The stamen generally consists of several thin stalks called **filaments**. At the top of each filament there is an **anther**. It is the anthers that produce the **pollen** grains.

FIG 5.2.1 The pistil and stamen of a *Hibiscus* flower

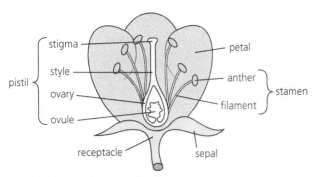

FIG 5.2.2 Parts of the pistil and stamen

The pistil consists of one, or sometimes several **styles**. At the top of each style there is a **stigma** where pollen grains become attached. The surface of the stigma is often sticky to increase the chances that a pollen grain will adhere to it. The bottom of the style is attached to the **ovary**, which contain one or more **ovules**.

Activity 5.2.1

Identifying the parts of the flower

Here is what you need:

- Scalpel/knife, hand lens, flowers from species such as Pride of Barbados (*Caesalpinia*), Poor man's orchid (*Bauhinia*) or Poinciana (*Delonix*).

1. Take a flower and examine it, turn it upside down. The green leaf-like structures at the base of the flower are the sepals. Remove them.

2. Next you will find the brightly coloured structures, these are the petals. Remove them. What do you notice about them? How many are present?

3. Inside of these you will find a ring of thin string-like structures. These are the stamens. At the top of the string-like structure (filament) is a round knob-like structure, the anther. The filament and anther together is the stamen, the male reproductive part of the flower.

4. Next is a swollen structure with a stalk on top of it. The swollen part is the ovary, the stalk is the style and on top of the style is the stigma. These structures together form the pistil, the female reproductive part of the flower.

5. Make a labelled drawing of each part of the flower that you have identified.

Fun fact

In some flowering plants the male and female parts of the flowers mature at different times. This prevents the stigmas from receiving pollen from the anthers of the same flower.

Check your understanding

1. Fig 5.2.3 shows the structure of a flower.

FIG 5.2.3

a) Identify the parts labelled w, x, y and z.

b) Which of these parts form the:
   i) pistil?
   ii) stamen?

Key terms

stamen male parts of a flower

pistil female parts of a flower

filament part of the male part of a flower that supports an anther

anther part of the male part of a flower where pollen is made

pollen male sex cell of a flowering plant

style part of the female part of a flower that connects the stigma to the ovary

stigma part of the female part of a flower that receives pollen

ovary part of the female part of a flower that contains ovules

ovule female sex cell of a flowering plant

Pollination

We are learning how to:

- describe the process of pollination
- identify the parts of a flower involved in pollination.

Pollination ⟩⟩⟩

For seeds to form, the ovules in an ovary must be fertilised by pollen grains. The transfer of pollen from the anthers of one flower to the stigma of the same flower or a different flower is called pollination.

- **Self-pollination** takes place when pollen is transferred to the stigma of the same flower or another flower on the same plant.

- **Cross pollination** takes place when pollen is transferred to the stigma of a flower on a different plant.

Pollination can take place either by transfer of pollen in the wind or on insects and other animals.

Pollination by animals

Many flowering plants are pollinated by insects or other animals like humming birds. The pollen grains are large and sticky so they can attach themselves to animals visiting the flowers.

Pollination by wind

Pollen of some flowering plants is carried by the wind. Plants that are wind pollinated produce pollen grains that are small, light and not sticky so they are easily carried through the air.

Wind pollinated flowers can often be identified by their structure.

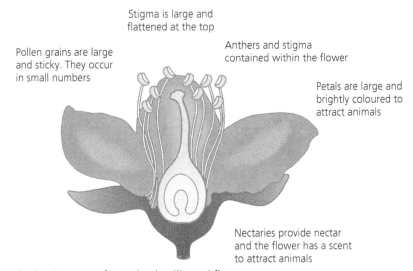

Stigma is large and flattened at the top

Pollen grains are large and sticky. They occur in small numbers

Anthers and stigma contained within the flower

Petals are large and brightly coloured to attract animals

Nectaries provide nectar and the flower has a scent to attract animals

FIG 5.3.1 Features of an animal pollinated flower

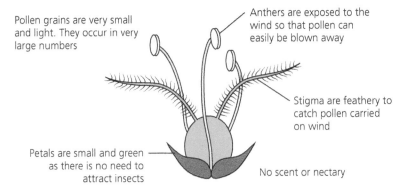

Pollen grains are very small and light. They occur in very large numbers

Anthers are exposed to the wind so that pollen can easily be blown away

Stigma are feathery to catch pollen carried on wind

Petals are small and green as there is no need to attract insects

No scent or nectary

FIG 5.3.2 Features of a wind pollinated flower

Comparing insect pollinated and wind pollinated flowers

Here is what you need:

- An example of a wind pollinated flower
- Hand lens.

Here is what you should do:

1. You have already identified the structures in the insect pollinated flower. Use the same techniques to identify the corresponding structures in the wind pollinated flower.

2. Are the structures similar in both? Account for any difference that might be found.

5.3

Fun fact

We most often associate insects with pollination but there are plenty of examples of flowers pollinated by other animals.

FIG 5.3.3 Humming birds pollinate some plants

The features of a wind pollinated flower and an animal pollinated flower are compared in Table 5.3.1. The features can be used to deduce the way in which an unknown flowering plant is pollinated.

Feature	Animal pollinated	Wind pollinated
Pollen	Small amount; large grains; sticky or covered in spikes	Large amount; small and light; not sticky
Petals	Large and brightly coloured; often scented	Small and dull coloured; not scented
Stamen	Short filaments; inside the flower	Long filaments; hang outside the flower
Stigma	Small and sometimes flat; sticky; inside the flower	Large and feathery; hangs outside the flower
Nectar	Present	Absent or very small amounts

TABLE 5.3.1

Check your understanding

1. FIG 5.3.4 shows the structure of a flower.

   Is this flowering plant more likely to be pollinated by the wind or by animals? Give reasons to support your answer.

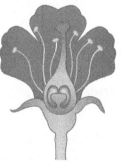

FIG 5.3.4

Key terms

self-pollination transfer of pollen on the same flower or another flower on the same plant

cross pollination transfer of pollen to the stigma of a flower on a different plant

Fertilisation and seed formation

We are learning how to:

- describe fertilisation in a flowering plant
- explain the formation of seeds.

Fertilisation ⟫

When a grain of pollen lands or is deposited on the stigma of a suitable flower, it must then combine with an ovule, but to do this it must reach the ovary, which is at the opposite end of the style.

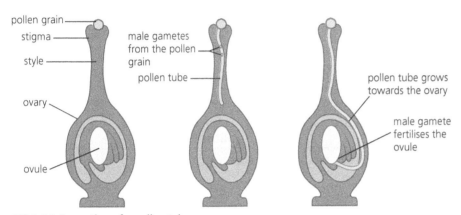

FIG 5.4.1 Formation of a pollen tube

To do this, a **pollen tube** grows down from the pollen grain, through the style, and into the ovary where it ends at an ovule.

Male **gametes** from the pollen grain can pass down the pollen tube and one will eventually fertilise the ovule to form a **zygote**.

Once the zygote is formed it will start to replicate, producing many cells, which differentiate and eventually form a seed.

FIG 5.4.2 Some fruits are succulent and edible

The pistil of a flower may be composed of one or more **carpels**. Each carpel will have a style and stigma, and an ovary. The ovules will become **seeds** while the remaining parts of the carpel will become the **fruit**.

In everyday language you use the term 'fruit' to describe something succulent and sweet that you eat.

However, in science any product from the fertilisation of a flower is termed a fruit.

Examining a longitudinal section of a flower

Here is what you need:

- Flower such as Poinciana, Poor man's orchid, Pride of Barbados or something similar
- Scalpel
- Hand lens.

Here is what you should do:

1. Remove the petals from the flower to leave the pistil and stamen.

2. Place the flower on a disecting board and carefully cut along its length.

3. Open out the two halves of the pistil and stamen.

4. Using a hand lens, carefully examine the inside of the flower, paying particluar attention to the ovary and the ovules.

5. Draw your specimen and label the ovary and ovules.

FIG 5.4.3 Ovary and ovules

FIG 5.4.4 The sweetest fruit in the world

FIG 5.4.5 Non-succulent fruits

There are many examples of non-succulent fruits.

Key terms

pollen tube tube through which male gametes from a pollen grain pass into the ovary and fertilise an ovule

gamete male or female sex cell

zygote formed when a male sex cell fertilises a female sex cell

carpel part of the pistil

seed formed from the ovule as a result of fertilisation

fruit formed, as a result of fertilisation, from the carpel excluding the ovules

Check your understanding

1. **a)** What is a pollen tube?

   b) What passes along a pollen tube?

2. What do each of the following become after fertilisation?

   a) The fertilised ovule

   b) The remaining parts of the carpel

Dispersal

We are learning how to:

- explain why seed dispersal is necessary
- describe different ways of seed dispersal.

The need to disperse seeds ⟩⟩⟩

Although seed formation is essential for each species of flowering plant to survive, this also provides the parent plant with a potential problem.

Plants rely on resources including light, water and nutrients from the soil in order to grow. If a plant simply dropped its seeds where it grew, there would soon be too many plants competing for the available resources.

As a result of overcrowding, no plants would flourish and they might all die off. In order to avoid this, plants have developed different ways of spreading or dispersing their seeds.

Seeds that are dispersed by **wind** are generally small and light.

Some seeds, like the poppy, rely on a good gust of wind to carry them away from the parent plant while others, like the dandelion, have parachutes or other structures that allow them to stay in the air a long time before they fall to the ground.

Seeds that are dispersed by **animals** rely on being eaten or carried.

Seeds which are eaten are at the centre of the fruit. When animals eat the fruits, like berries, the seeds pass through the animal's body undamaged and are egested several hours later at a location some distance from the parent plant.

Fruits like those of the agrimony plant are not good to eat but they are covered in small hooks that attach the fruit to passing animals. The fruits will eventually be rubbed off or fall off a long way from the parent plant.

The coconut is a huge seed. It is far too big to be dispersed by wind or animals.

The coconut floats in **water** and is dispersed by being carried by moving water.

FIG 5.5.1 Wind dispersed seeds: **a)** poppy seeds **b)** dandelion seeds

FIG 5.5.2 Animal dispersed seeds: **a)** hawthorn berries **b)** agrimony fruits

FIG 5.5.3 The coconut is dispersed by water

Some plants can disperse their own seeds and do not rely on an outside agent such as the wind or an animal.

This is possible by the process of **dehiscence**. When the seed pods of some flowering plants mature they start to dry out and distort. Different parts of the pod are under tension. which is suddenly released when the seed pod springs open. This is enough to flick the seeds a considerable distance from the parent plant.

FIG 5.5.4 The acacia disperses its own seeds

Activity 5.5.1

Examining seeds and fruits

Here is what you need:

- Gather a number of seeds and fruits from local plants
- Hand lens.

Here is what you should do:

1. Examine each seed / fruit carefully. Use the hand lens to see detail more clearly.
2. Decide how the plant disperses its seeds.
3. Draw up a table and make a list of plants which disperse their seeds in the different ways described.

Check your understanding

1. Copy and complete the following table to summarise the differences between seeds that are dispersed by the wind and those that are dispersed by animals.

	Wind dispersed	Animal dispersed
Size		
Weight		
Seed contained in a juicy fruit?		
Fruit covered in hooks or hairs?		

TABLE 5.5.1

> **Fun fact**
>
> Elephants are thought to be the main dispersers of seeds in many jungles. They eat plants containing seeds in one place and may deposit the seeds in egested food several kilometres away, where they will germinate. The egested food provides the developing seedlings with nutrients.

Key terms

wind dispersal seeds carried away by the wind

animal dispersal fruits eaten or attached to the coats of animals

water dispersal seed carried in moving water

dehiscence seed pods explode and flick seeds away

Germination

We are learning how to:

- state the different conditions necessary for the germination of seeds
- describe what happens immediately after germination.

Germination >>>

A seed contains all that is necessary to grow into a new plant. When conditions are suitable a seed will **germinate** and start to develop.

The tough outer coating, or **testa**, is the most obvious part of a seed you observe. Closer examination reveals the **hilum**, which is a scar showing where the seed was attached to the ovary during its development. You may also be able to see the **micropyle**. This is a tiny hole in the testa through which water can enter the seed.

Inside the seed are the **cotyledons**. These are the food store of the seed and will provide the germinating plant with all the nutrients it needs until it can start to make food for itself. In some plants, the cotyledons will become the first leaves.

Sandwiched between the cotyledons are the **radicle**, which will develop into the root of the plant, and the **plumule**, which will develop into everything above the ground, such as the stem, leaves and flowers. In germinating seeds, the radicle grows first, followed by the plumule.

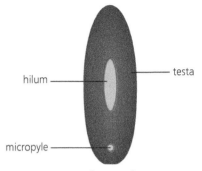

FIG 5.6.1 External parts of a germinating seed

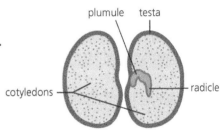

FIG 5.6.2 Internal parts of a germinating seed

Activity 5.6.1

Structure of a germinating seed

Here is what you need:

- Bean seed that has been pre-soaked so it is just starting to germinate
- Scalpel
- Hand lens.

Here is what you should do:

1. Examine the bean seed and identify the testa and the hilum.
2. Gently squeeze the seed and identify the micropyle where you see a drop of water forming.
3. Use the scalpel to gentle open the bean seed down the middle and carefully examine it.
4. Identify the plumule, the radicle and the cotyledons.
5. Make a drawing of your specimen and label the parts mentioned.

> **Fun fact**
>
> The viability of seeds decreases with age. In general, the older seeds are the lower the percentage that will germinate. However, in harsh environments, like deserts, where conditions for germination occur irregularly, seeds may remain viable in the ground for many years.

Conditions needed for germination

For seeds to germinate they must have:

- water or moisture

- oxygen from the air – this is needed for respiration to provide energy

- warmth.

Seeds don't need light to germinate. We know this because we often sow the seeds on the surface of soil and cover them to prevent birds from eating them. Light is only needed by seedlings once they have some green leaves and can carry out photosynthesis to make food.

Activity 5.6.2

Germinating bean seeds

Here is what you need:

- Beaker
- Absorbent paper
- Bean seeds × 4.

Here is what you should do:

FIG 5.6.3

1. Dampen the absorbent paper with water.

2. Place the absorbent paper into the beaker so it fits inside the beaker and pushes against the side of the beaker.

3. Place four bean seeds between the beaker and the absorbent paper. The beans should be in the different orientations shown in Fig 5.6.3.

4. Put some water into the bottom of the beaker and leave it for several days until the beans start to germinate (Fig 5.6.4). Don't let the beaker dry out.

5. When the roots and shoots are growing on each seed, draw the germinating seeds.

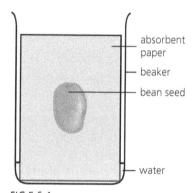

FIG 5.6.4

absorbent paper
beaker
bean seed
water

Check your understanding

1. **a)** Identify the parts A to D on the seed shown in Fig 5.6.5.

   b) What conditions does a seed need to germinate?

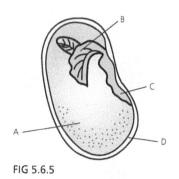

FIG 5.6.5

Asexual reproduction in plants

We are learning how to:

- explain the difference between sexual and asexual reproduction in plants
- describe different ways in which plants may reproduce asexually.

Sexual reproduction »

The processes of pollination, fertilisation, seed formation and germination are all part of the cycle of **sexual reproduction** in plants. Two parents are required: one that provides the male sex cells or pollen, and one that provides the female sex cells or ovules.

During self-pollination both sex cells may be provided by the same flower or by different flowers from the same plant, but they are nevertheless considered as different parents. The different sex cells provide a mixture of inherited features so the plants which grow from the resulting seeds are unlikely to be identical to the parent plant(s)

Asexual reproduction

Some plants are able to reproduce without forming seeds. This may happen in a variety of ways but they all have one thing in common. The offspring come from a single parent plant.

This process is called **asexual reproduction or vegetative reproduction.** Since the inherited features are passed on by a single parent, the offspring plant will be identical to the parent plant in every way and is described as a **clone** of the parent.

When bulbs or tubers are planted in the ground they will produce **clumps** of plants over time. Each plant in the clump will be a clone of the parent plant.

Some plants send out stems which grow on the surface of the soil or just below it. These are called **stolons** or **runners**. At the end of each runner a new plant will form. It will be a clone of the parent plant.

FIG 5.7.1 Clumping

FIG 5.7.2 Stolons or runners

FIG 5.7.3 Layering

The branches of some plants grow close to the ground as they get longer. Eventually the end may lay on the soil and roots will form. Over time a new plant forms and becomes separate from the parent plant. This method of vegetative reproduction is called layering. Once again, the offspring is a clone of the parent plant.

Activity 5.7.1

Examining an example of vegetative reproduction

Here is what you need:

- Your teacher will provide you with a parent plant which is producing offspring by vegetative reproduction
- Hand lens
- Plant pot
- Compost.

Here is what you should do:

1. Carefully examine the parent and the offspring. Use the hand lens to see the fine detail.

2. Satisfy yourself that the offspring is a miniature version of the parent plant.

3. Put some compost into a plant pot.

4. Separate one or more offspring from the parent plant and place it in the compost.

5. Place the plant pot somewhere light but not in direct sunlight.

6. Water the plant regularly and, if possible, take a photograph of it every couple of days so you can show how it develops.

Check your understanding

1. FIG 5.7.5 shows the plant *Chlorophytum comosum*. It is sometimes called the 'spider plant'.

   a) How is the plant producing offspring?

   b) How do the features of the offspring compare to those of the parent plant?

   c) How could the offspring be grown into mature plants?

FIG 5.7.5 *Chlorophytum comosum*

Fun fact

The plant *Bryophyllum daigremontianum* is also known as the 'Mother of thousands' plant. It has lost the ability to produce seeds but tiny plants develop along the edges of its leaves.

FIG 5.7.4 *Bryophyllum daigremontianum*

Key terms

sexual reproduction reproduction involving two parents

asexual reproduction reproduction involving one parent

clone offspring which has identical features to a parent plant

clump group of plants formed from one parent

stolon stem that grows on or just below the soil surface and produces new plants

runner another name for a stolon

Commercial importance of vegetative reproduction

We are learning how to:

- explain the commercial importance of different forms of vegetative reproduction.

Commercial importance of vegetative reproduction

In the previous lesson you saw how some plants naturally undergo vegetative reproduction. There are certain advantages to horticulturalists of producing plants in this way:

- Plants will grow to maturity much more quickly than seeds.

- Offspring will share any desirable features of the parent plant.

- It provides an easier option when seeds are scarce or difficult to germinate.

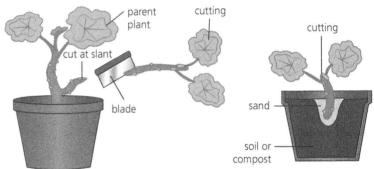

FIG 5.8.1 Taking a cutting

You have already learned about **layering**. Plants can be encouraged to produce offspring by pegging the ends of branches into the soil. One parent plant might produce several identical offspring at the same time.

A **cutting** is a shoot cut most often from a plant stem. It has leaves but no roots. Some cuttings will develop roots if suspended in water, and can then be planted. Other cuttings may be placed directly into compost. Dipping the end of the cutting in hormone rooting compound helps root growth.

FIG 5.8.2 Banana offsets

Bananas do not produce seeds. When the parent plant has flowered and fruited it will die but before it dies it will produce **offsets** around its roots, which can be separated. These offsets rapidly develop into mature plants.

Grafting provides a method of attaching the stem and leaves of a plant onto a different root. There are a number of different sorts of grafting, such as cleft grafting (Fig 5.8.3).

One use of grafting is to grow fruit trees on dwarf roots stocks. The scion is a fruit tree while the stock is the root from another dwarf tree. This limits the height of the tree so the fruit can be grown in smaller areas and harvested more easily.

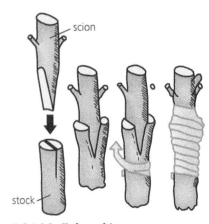

FIG 5.8.3 Cleft grafting

Activity 5.8.1

Taking a cutting of a geranium plant

Here is what you need:

- Joseph's coat (*Coleus* spp.) or Dragon's blood plant
- Scalpel or razor blade
- Beaker or clear plastic cup.

Here is what you should do:

1. Cut a length from one of the stems of your geranium. This should be around 10–15 cm and have a few leaves. The cut should be clean and not ragged.

FIG 5.8.5

2. Place the cutting in a beaker of water and leave it somewhere light but not in direct sunlight.

3. Check the water level each day to ensure that the container doesn't dry out.

4. Record how many days it takes for roots to appear.

5. Take pictures at different times of the root development.

6. When there is sufficient root development plant your cutting in a container of soil or compost.

7. Compare the appearance of the cutting with the parent plant. For example, are the leaves the same shape and colour?

Fun fact

Horticulturalists are able to produce fruit trees, such as apple trees, that produce two or more different varieties of apple by grafting scions from different trees onto the same root stock.

FIG 5.8.4 Two varieties of apple on the same tree

Key terms

layering producing a new plant by pegging a stem of the parent plant in the soil

cutting producing a new plant from part of the stem of a parent plant

offset small growth from the root of a plant

grafting combining parts of two different plants

Check your understanding

1. Explain the difference between cutting and grafting.

2. Why are offsets important in the banana fruit growing business?

Review of Plant reproduction

- The life cycle of a flowering plant involves germination, pollination, fertilisation, seed formation and seed dispersal.

- The flower is the organ of sexual reproduction in a flowering plant.

- A flower consists of a male part, the stamen, and a female part, the pistil.

- The stamen consists of anthers and filaments.

- The pistil consists of the stigma, style, ovary and ovules.

- Pollination is the transfer of pollen from the anthers of a flower to the stigma of the same or a different flower.

- Pollination is most often carried out by animals or the wind.

- Flowers which are wind pollinated are different in structure from those which are animal pollinated.

- Fertilisation occurs when a male gamete from a pollen grain passes along the style of a flower into the ovary where it combines with an ovule to form a zygote.

- The fertilised ovules of a flower will eventually become seeds.

- The remaining parts of the carpel will become the fruit.

- A plant disperses its seeds to avoid competition for light, water and nutrients.

- Seeds are often dispersed by wind and animals, and less often by dehiscing and water.

- A seed consists of a food store in the form or one or two cotyledons, surrounded by a protective coat called the testa.

- Germination is the period during which a seed begins to grow.

- A small hole in the testa called the micropyle allows water to enter the seed.

- When a seed germinates, the root of plumule grows first and then the shoot or radicle.

- A flowering plant may also reproduce asexually; this is sometimes called vegetative reproduction.

- Examples of vegetative reproduction include bulbs, tubers and runners.

- A plant produced by asexual reproduction is an exact replica or clone of the parent plant.

- Vegetative reproduction is important in commercial plant production; layering, cuttings, grafting and offsets are used to increase plant numbers.

Review questions on Plant reproduction

1. Here are five processes that take place in the life cycle of a flowering plant.

    dispersal   germination   growth   pollination   seed formation

    Write them in order starting with germination.

2. Briefly describe what happens during the following processes.

    a) Pollination   **b)** Germination   **c)** Fertilisation   **d)** Dispersal

3. The diagram shows six parts (A to F) of a flower.

    a) In which part is pollen made?

    b) Which parts form the female part of the flower?

    c) Which part is coloured to attract insects?

    d) In which part does fertilisation take place?

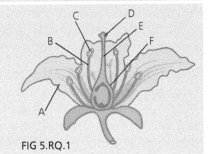

FIG 5.RQ.1

4. Copy and complete the following table which compares the structure of a wind pollinated flower with an animal pollinated flower.

Feature	Animal pollinated	Wind pollinated
Pollen	Large grains, sticky or spiky, small amount	
Petals	Brightly coloured, big	
Stamen	Filaments short, inside flower	
Stigma	Small, sticky; inside flower	

TABLE 5.RQ.1

5. **a)** Draw a diagram to show how the pollen which lands on the stigma is able to fertilise an ovule in the ovary.

    b) What is a zygote?

    c) What does the fertilised ovule eventually become?

6. The following photographs show the fruits of two plants, A and B.

FIG 5.RQ.2

For each plant, determine whether the seeds are dispersed by wind or by animals and explain your answer.

Propagating plants from cuttings

Mr Patterson is planning to start a business selling flowering plants. His plan is to produce large numbers of plants for sale in plant centres.

He understands about obtaining plants by growing from seeds but he also wants to propagate plants from cuttings and he is uncertain about the best way to go about this.

FIG 5.SIP.1 A plant centre

Mr Patterson carried out some research at local gardening shops but this made him even more confused. He found that there are a number of products on the market to help him root his cuttings. They all claim to help cuttings to develop roots but which one is best? Will they all produce the same result, in which case he might as well buy the cheapest, or is one product better than the others, in which case it might be worth spending a little more?

Mr Patterson is investing a lot of money into his business so it is important that he finds out which of the rooting products is going to give him the best results. Before he starts taking cuttings on a large scale he has hired you to investigate and make recommendations.

FIG 5.SIP.2 Rooting hormone

1. You are going to work in groups of 3 or 4 to investigate which rooting hormone product works best on plant cuttings. The tasks are:

   - To review how to propagate a plant by taking a cutting.
   - To devise a standard method of taking cuttings.
   - To carry out research into what products which promote cuttings to grow roots are available in your local gardening shops.
   - To make a note of the active chemical(s) in each product and read the instructions for how the product should be used.
   - To plan a fair test in which you will compare the performance of each product.
   - To decide how you will evaluate the results of your test in order to make recommendations.
   - To compile a report, including a PowerPoint presentation in which you explain how you carried out your fair test, show photographs of the results and present the conclusions you have come to about the different products tested.

a) Look back through the part of the unit that describes asexual reproduction in flowering plants and make sure that you understand the principles of propagation from cuttings.

b) Devise a standard method of taking cuttings. You need to consider such factors as:

- Should I take the cutting from new green growth or old brown growth?
- What thickness of stem should I take?
- Should I cut straight across the stem or on a slant?
- How many leaves should I leave on the cutting?

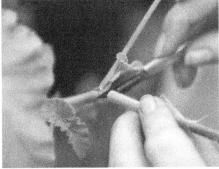

FIG 5.SIP.3 Taking a cutting

c) Carry out research into what rooting aids are available in your local garden shops and obtain a sample of each. How much is each product? Is the active chemical(s) identified on the label? Read the instructions for using each product.

d) Devise a fair test to compare the rooting aids. You need to consider:

- Which type of plant will provide the cuttings?
- How many cuttings will I use with each product?
- What conditions do I need to keep the same in order to make it a fair test e.g. size of cutting, amount of compost, amount of water given, light intensity, temperature?

e) How are you going to decide how well or how badly each product performed? For example, you might take pictures of the developing roots of each sample every few days. If the pictures for each sample were placed in a line you would have a visual record of how the roots developed.

How will you compare the performance of each product? For example, you might measure the number of days taken for roots to develop to a certain length or you might compare root density after so many days.

Is there any way you can relate the performance of the rooting aids to their cost? For example, if one rooting aid cost twice as much as another did it perform twice as well?

f) You will have carried out your test on cuttings from one type of plant. If time allows you might test the rooting aids that gave best results on other types of plants to see if they give good overall results.

g) Prepare a PowerPoint presentation in which you describe what you did by way of carrying out a fair test.

Show the data obtained, in the form of photographs and other observations and measurements you may have taken.

Make your recommendations to Mr Patterson based on your investigation.

Unit 6: Sexual maturity, reproduction and personal hygiene

We are learning how to:

- recognise sexual maturity
- describe reproduction
- appreciate the importance of personal hygiene.

Growth is one of the several characteristics of all living things. One of the features of human growth is that our body increases in size until we become adults around the age of 20 years old.

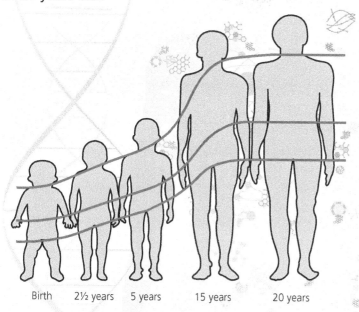

Birth   2½ years   5 years   15 years   20 years

FIG 6.1.1 Human growth pattern

Growth is not just about getting bigger. As a person grows their bodies undergo other physical changes and chemical changes that affect the way we behave.

We can conveniently divide the human life cycle into four stages:

infancy ⟶ childhood ⟶ adolescence ⟶ adulthood

Once humans reach adulthood they can reproduce. This is another characteristic of all living things.

In humans, sexual reproduction takes place between a man and a woman. This requires specialised reproductive systems which only start to function as we grow.

FIG 6.1.2 Man and woman

The male sex cells are called sperm and the female sex cells are called ova (singular ovum). For reproduction to occur the sperm must leave the male body and enter the female body where it joins an ovum and fertilisation takes place. This results in the formation of an embryo, which grows inside the woman.

To look after the body at different stages of our lives, we must learn to practice personal hygiene.

When we are infants and children our parents take responsibility for our personal hygiene. They make sure we are washed regularly and learned to carry out self-maintenance like cleaning our teeth, trimming our nails and brushing our hair.

As we grow older we learn how to look after ourselves. We take on responsibility for our own personal hygiene.

FIG 6.1.3 When we are young our parents look after our personal hygiene

FIG 6.1.4 As we grow we learn about personal hygiene

Infancy and childhood

We are learning how to:

- describe the different stages of the human life cycle
- outline how we develop during infancy and childhood.

A new-born baby is born it is totally dependent on its parents and other adults to protect it and feed it. Children remain dependent on adults until they are well into their mid-teens.

Childhood is the time when children are also protected from diseases by **vaccination**.

Jamaica has an excellent immunisation programme. As a result, it has been possible to eliminate diseases like poliomyelitis, measles and rubella. Children are routinely vaccinated against poliomyelitis, measles, mumps, rubella, diphtheria, tetanus, pertussis, hepatitis and haemophilus influenza.

For a child to develop correctly, he or she must receive all of the nutrients needed to remain healthy in the correct amounts. If they don't they will be **malnourished**. A child suffering from **anaemia** may receive plenty of food but is malnourished because there is a lack of iron in their diet.

FIG 6.2.1 Infancy is the first stage of our development

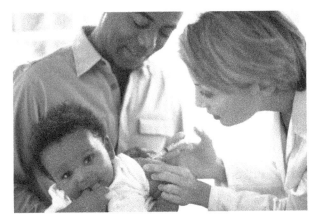

FIG 6.2.2 Vaccination protects children from disease

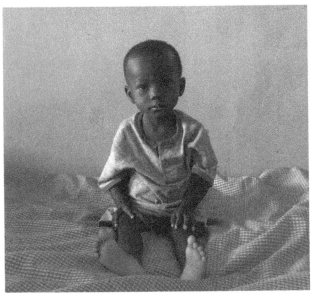

FIG 6.2.3 A malnourished child

FIG 6.2.4 How many heads high are you?

There is a lot more to human growth than simply becoming taller or heavier. As a child grows, their body changes shape. For example, the body grows more than the head. An infant may be 4 times the height of his or her head while an adult may be 7½ times. The values given in Fig 6.2.4 are only approximate because people are all different.

Activity 6.2.1

How do children change as they grow?

You should work in small groups for this activity.

As babies develop into children they not only grow but they gradually learn how to do things for themselves. Discuss how babies change in appearance as they grow and the things they learn to do. Write two lists in the form of a table like the one below.

Ways in which a ten-year old looks different from a baby	Things a ten-year old can do that a baby cannot do

TABLE 6.2.1

Check your understanding

1. Why is a baby totally dependent on its parents and others?

2. **a)** Why are babies vaccinated?

   b) Name three diseases against which babies are vaccinated.

3. Why is it important that children receive the right kinds of food rather than large quantities of food?

Fun fact

Some people mistakenly think that being malnourished means not receiving enough food. What it really means is not receiving the nutrients needed by the body in the correct amounts. A person who is obese is malnourished because they receive more nutrients than their body requires.

Key terms

vaccination treatment with a vaccine to prevent catching a disease

malnourished not receiving all of the nutrients needed for healthy growth in the correct amounts

anaemia disease caused by insufficient iron in the diet

Adolescence and adulthood

We are learning how to:

- describe different stages of the human life cycle
- outline how we develop during adolescence and adulthood.

In general, the rate that a child grows gradually decreases as they get older. However, the decrease is interrupted shortly before the end of the growth period. At this time there is marked acceleration of growth, called the **adolescent growth spurt**. In boys this takes place from about 13 to 15 years, and in girls from about 12 to 14 years.

Adolescence is a period of life during which we are no longer children but not yet adults. Many changes take place to our bodies, and to the way in which we think about things. The start of adolescence is marked by **puberty**. This is the time when the male testes start to make sperm and the female ovaries start to make eggs.

Both boys and girls are born with a complete set of sex organs but they only become active as they grow. When we reach adolescence the **pituitary gland** at the base of the brain starts to make chemical messengers called **hormones**. When these are released into the bloodstream, they activate the sex organs which, in turn, start producing sex hormones.

FIG 6.3.1 A person must come to terms with their changing body during adolescence

The sex hormones are responsible for changes that take place to the body during adolescence. These changes are sometimes described as **secondary sexual characteristics**.

The hormone released by the male sex organs is called testosterone. As well as stimulating the production of sperm in the testes, this hormone also causes:

- hair to start to grow on the face and body, the arm pits and the pubic area
- the voice to deepen
- the body to become more muscular.

The hormone released by the female sex organs is called oestrogen. As well as stimulating the production of eggs, this hormone also causes:

- hair to grow on parts of the body; mainly the arm pits and the pubic area
- the breasts to develop
- the hips to widen
- periods to start.

Adolescence can be a difficult and sometimes emotional time for a person. The release of hormones associated with puberty can bring about mood swings and a chemical imbalance leading to conditions like acne.

A person must also come to terms with their changing body. Adolescence is a time when people start to think in a more adult way about things and experience increased sexual urges. Boys and girls start to appreciate that they are different and that they are attracted to each other.

Adolescence generally finishes around the age of 18 years. The subsequent change to adulthood takes place over several years as an individual takes on their adult form and their behaviour becomes more mature. Part of being an adult is taking on responsibilities for yourself and those around you.

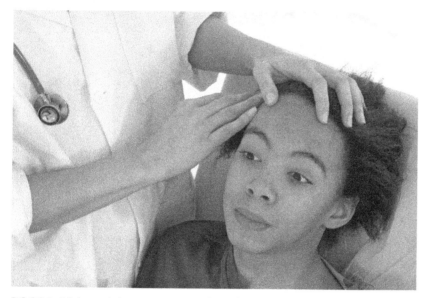

FIG 6.3.2 Adolescents become very conscious about their appearance

Key terms

adolescent transition period between childhood and adulthood

puberty time when the body becomes sexually mature

pituitary gland part of the brain that releases hormones into the blood stream

hormone chemical messenger releavsed in one part of the body, carried in the blood and brings about change in another part of the body

secondary sexual characteristics physical characteristics developing at puberty which distinguish between the sexes

Check your understanding

1. Arrange the following in the order they take place.

adolescence   adulthood   childhood   infancy

2. a) What is a hormone?

b) Name a hormone involved in the development of secondary sexual characteristics in:

i) a girl   **ii)** a boy.

How are we different?

We are learning how to:
- describe different stages of the human life cycle
- appreciate that people develop in different ways at different rates.

Although boys and girls appear to grow at the same rate this is not so.

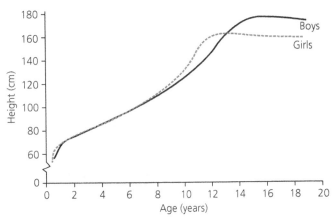

FIG 6.4.1 How the heights of boys and girls change from birth to adulthood

Although boys generally grow taller than girls, girls reach their final height at a younger age. However, charts like the one in Fig 6.4.1 are always based on **average** values because people of the same age show **variation**.

Students in a class often show differences in height and body mass. Some students are taller than others while others are heavier than others. This doesn't mean there is anything wrong with any of them but it simply reflects that people are all a little different.

The size of a person is determined by factors like:

- the size of their parents. Large parents are likely to have large children

- the speed with which they grow. Some people grow more quickly than others and reach their adult size sooner.

FIG 6.4.2 Students in the same class are often different heights

FIG 6.4.3 Students in the same class often have different body mass

Activity 6.4.1

Variation in the height of students in the class

Your teacher will help you to organise this activity.

Here is what you need:

- Measuring tape for this activity.

Here is what you should do:

1. Measure the height of each student in the class to the nearest centimetre.

2. Write the value on the board so that everyone can copy this down into their exercise book.

3. Draw a frequency table like the one below and complete it using your results. Start from the smallest height in your class.

Height in cm	Number of students
140–144	
145–149	

TABLE 6.4.1

4. Use your results to draw a histogram showing the heights of the students in your class.

Check your understanding

1. The following table contains data about the body mass of 40 male students in a class.

Body mass in kg	Number of students
41–44	4
45–50	7
51–55	9
56–60	11
61–65	6
66–70	3

TABLE 6.4.2

a) What word is used to describe differences in features like body mass within a group?

b) Suggest two reasons why the students do not all have the same body mass.

c) Draw a histogram to represent this data.

d) How would a histogram of 40 female students of the same age be:

i) similar? ii) different?

Fun fact

Jamaican basketball player Marvadene Anderson was 6 feet 11 inches (2 m 11 cm) tall when she was only 16 years old.

FIG 6.4.4 Guess which player is Marvadene Anderson?

Key terms

average typical value for a set of data

variation difference in a feature such as height or body mass

The male reproductive system

We are learning how to:

- describe different stages of the human life cycle
- identify the parts of the male reproductive system and their functions.

The male reproductive system ≫

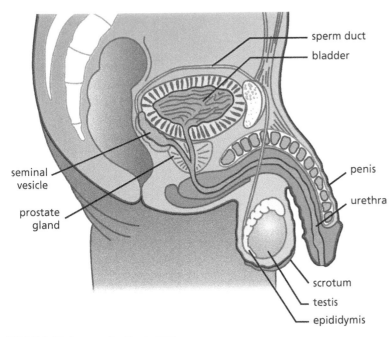

FIG 6.5.1 Male reproductive system

The male reproductive organs are the two **testes**. These lie outside the body cavity in a sac called the **scrotum**. This allows the testes to remain at a temperature that is slightly below normal body temperature. This favours the production of **sperm**.

Each testis contains many tubes in which sperm is formed. These meet and join to connect with the epididymis.

The epididymis leads to the **sperm duct**. The two sperm ducts open into the ureter just after it leaves the bladder. Urine from the bladder and sperm both pass out of the **penis** through the urethra. The body has a mechanism that prevents these events happening at the same time.

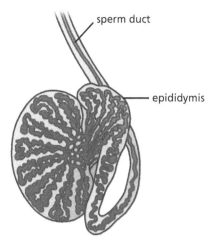

FIG 6.5.2 Structure of a testis

sperm duct

epididymis

The seminal vesicle branches from each sperm duct just before it enters the prostate gland. The seminal vesicle contains fluid that mixes with the sperm to form semen. The prostate gland secretes a fluid that nourishes the sperm. Millions of sperm are released each time a male ejaculates.

Activity 6.5.1

Tracing the movement of sperm

Here is what you need:

- Model of the male reproductive organs (if this is not available use Fig 6.5.1).

Here is what you should do:

1. Follow the passage of the sperm from where it is formed to where it leaves the body on the model.

2. Make a list of the parts of the male reproductive system in the order that sperm passes through them.

Check your understanding

1. In which parts of the male reproductive system is sperm produced?

2. What is the function of the seminal vesicles?

3. What is the name of the duct that joins the epididymis to the urethra?

4. What else apart from sperm leaves the body through the urethra?

Fun fact

A vasectomy is a procedure a man can have if he and his partner agree that they do not want to have any more children. It involves a minor operation during which a short section of each sperm duct is removed and the remaining ends tied off. This prevents sperm from passing from the testes to the urethra. A vasectomy does not affect a man's ability to have sexual intercourse.

Key terms

testes male reproductive organs

scrotum sac outside the body that contains the testes

sperm specialised reproductive cells produced in the testes

sperm duct tube along which sperm pass before they reach the prostate gland

penis part of male reproductive system through which sperm and urine pass

The female reproductive system

We are learning how to:

- describe different stages of the human life cycle
- identify the parts of the female reproductive system and their functions.

The female reproductive system ⟫

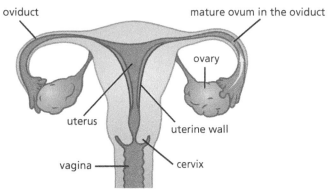

FIG 6.6.1 All of the parts of the female system are held within the body cavity

The female reproductive organs are two **ovaries**. These are found at the back of the abdomen, just below the kidneys. **Ova** develop in the ovaries.

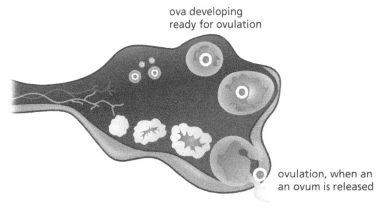

FIG 6.6.2 Structure of an ovary

When a female is mature, her ovaries will release one ovum each month. This is called **ovulation**.

Fun fact

Men produce new sperm daily throughout their lives. Women are born with all of their ova. These are in the form of immature ova or follicles, which are stored in the ovaries.

A woman will have between one and two million follicles at birth. Of these, only about 400 will mature during the woman's child-conceiving years, while the rest will die.

When a woman can no longer have children, few or no follicles will remain in her ovaries.

Close to each ovary is the funnel-shaped opening of an oviduct. The oviducts are also called **fallopian tubes**. They are narrow tubes along which the ova pass from the ovaries to the **uterus**. Fertilisation of an ovum by a sperm normally happens in the oviduct.

The uterus is wider than the oviducts. It is the place where a fertilised ovum will develop into an embryo and eventually into a baby.

The uterus, which is usually about 80 mm long, connects with the outside of the body via a muscular tube called the **vagina**. The neck of the uterus is a ring of muscle called the **cervix**.

The urethra, which carries urine from the bladder, opens at the outer end of the vagina.

Activity 6.6.1

Tracing the movement of ova

Here is what you need:

- Model of the female reproductive organs (if this is not available use Fig 6.6.1).

Here is what you should do:

1. Follow the passage of the ova from where they are formed to where they leave the body (assuming they are not fertilised) on the model.

2. Make a list of the parts of the female reproductive system in the order that the ova pass through them.

Check your understanding

1. In which parts of the female reproductive system are ova produced?

2. What is the name of the duct that joins an ovary to the uterus?

3. What is the cervix and where is it found?

4. How many ova are normally released during ovulation?

Key terms

ovaries female reproductive organs

ova specialised female reproductive cells

ovulation when a mature female releases one ovum each month

fallopian tubes narrow tubes along which the ova pass from the ovaries to the uterus

uterus the place where a fertilised ovum will develop into an embryo and eventually into a baby

vagina muscular tube connecting the uterus with the outside of the body

cervix ring of muscle where the uterus joins the vagina

The menstrual cycle

We are learning how to:

- describe different stages of the human life cycle
- explain what happens to the female body during the menstrual cycle.

When a woman reaches an age when she can have children, she has a regular **menstrual cycle**. This typically lasts for 28 days but varies from woman to woman.

Women start having menstrual cycles during puberty at around 12 years old and continue to have them until the menopause at around 50 years of age. During a menstrual cycle, there are changes to the level of female hormones in the body and changes in the ovaries and uterus.

The cycle starts when the follicle stimulating hormone (FSH) is released from the pituitary gland in the brain. This causes an egg to mature in a **Graafian follicle** inside one or other of the two ovaries.

| ovum matures inside Graafian follicle | duration day 14 | corpus luteum develops | corpus luteum degenerates | |

Events occurring in the ovary

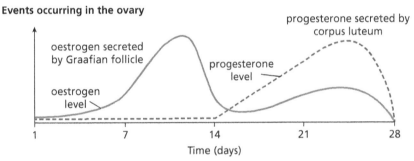

(a) The menstrual cycle

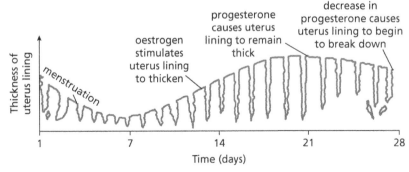

(b) Hormones in the menstrual cycle

FIG 6.7.1 Changes take place in the body during the menstrual cycle

The Graafian follicle also secrets a hormone called oestrogen. This hormone causes the lining of the uterus to thicken. If fertilisation occurs during the cycle, the fertilised egg will become embedded in the thickened uterus wall and develop into a baby.

As the level of oestrogen begins to drop, the pituitary gland releases another hormone called luteinising hormone (LH) that triggers ovulation. The mature egg is released by the ovary.

The remains of the Graffian follicle form another structure called the **corpus luteum**. About a week after the egg is released the corpus luteum starts to produce a hormone called progesterone. If the egg is not fertilised the corpus luteum disintegrates and the level of progesterone falls. The lining of the uterus breaks down and is discharged from the body through the vagina. This process is called **menstruation**, or having a period.

The first day of menstruation is taken as the first day of the menstrual cycle because this is the day most easy to identify. Ovulation occurs around about the 14th day of the cycle. Sexual intercourse in the days immediately after ovulation is most likely to result in fertilisation and pregnancy.

Fun fact

Corpus luteum is Latin for body of yellow or yellow body.

Activity 6.7.1

Teenage pregnancy
There are a number of reasons why a woman should wait until she is an adult before having a baby. Most of these apply equally well to a man. They are to do with issues like:
- the physical development of the woman
- the cost
- the need for maturity and patience
- providing a good role model.
Here is what you should do:

In your groups, discuss why it is better for a man and a woman to wait until they are older and in a steady relationship before having a baby.

Share your findings with other members of the class.

Key terms

menstrual cycle regular cycle of egg release by a female between the ages of about 12–50 years

Graafian follicle part of the ovary from which eggs are released

corpus luteum remains of the Graafian follicle after an egg is released

menstruation period when blood is lost during the menstrual cycle

Check your understanding

1. **a)** In which structure in the ovary does an egg mature?
   b) Which hormone causes this to take place?
   c) Which hormone triggers the release of an egg from an ovary?
   d) What happens to this structure when the egg is released?
2. **a)** Which hormone causes the uterus wall to thicken?
   b) Why does this happen?
   c) What happens to the thickening of the uterus wall if the egg is not fertilised?

Sexual reproduction

We are learning how to:

- describe different stages of the human life cycle
- explain what happens during sexual reproduction.

Fertilisation in humans is the coming together of a sperm or male gamete and an egg or female gamete to form a **zygote**. It is from the zygote that the **embryo** will form and develop to become a new-born baby.

Fertilisation takes place in the body of the female.

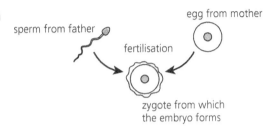

FIG 6.8.1 Fertilisation

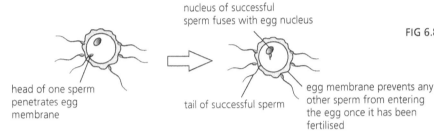

FIG 6.8.2 An egg is only fertilised by one sperm

An egg cell is only fertilised by a single sperm cell. Once a sperm has entered the egg rapid changes take place to the membrane surrounding the egg. This prevents other sperm from entering.

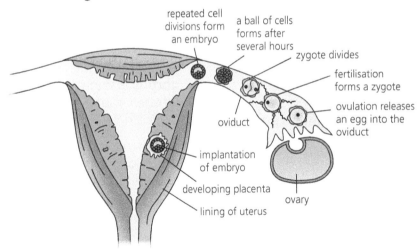

FIG 6.8.3 Fertilisation usually takes place in the oviduct

Sperm is deposited in the vagina of the female from where it swims towards the ovaries. Fertilisation usually takes place in the woman's oviduct as the egg is passing down from the ovary. Soon after the zygote forms it starts to divide and this process continues as more and more cells are formed. At this time the zygote becomes an embryo.

Fun fact

The oviducts are also called the Fallopian tubes

In this part of the woman's menstrual cycle, her uterus wall has become thickened ready to receive the embryo. As the embryo enters the uterus it sinks into the wall in a process called **implantation**.

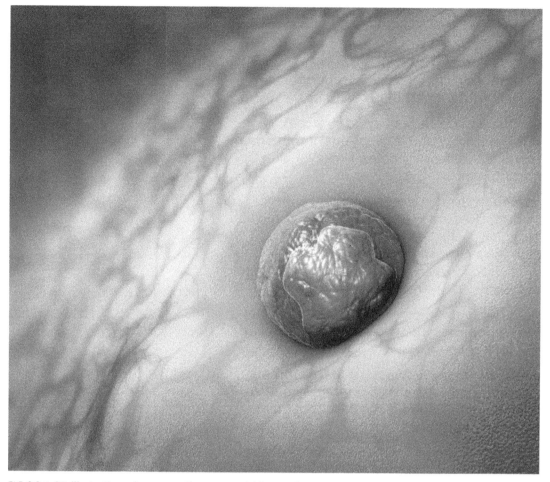

FIG 6.8.4 3D illustration of an egg cell or ovum sticking to the uterus.

Check your understanding

1. Where does fertilisation normally take place?
2. When does an embryo become a foetus?
3. Why is it that an egg can only be fertilised by one sperm?
4. What happens to the zygote shortly after it is formed?

Key terms

fertilisation coming together of the male and female sex cells

zygote formed by the combination of a male sex cell and a female sex cell

embryo develops from a zygote

implantation process in which the embryo attaches to the uterus wall

Personal hygiene

We are learning how to:

- describe different stages of the human life cycle
- ensure our own personal hygiene.

By the time a person becomes an adolescent they should have taken over responsibility for their own personal hygiene. This is an important part of growing up.

Skin is a remarkable material. It bends and stretches as we move about, it is waterproof and it keeps out harmful germs that would otherwise invade our bodies. However, skin needs regular maintenance.

Skin gets dirty through the day as we come into contact with grime and dust. Urea is a waste product that is released by the body. Most leaves the body in urine but some is excreted through the skin as sweat. If we don't wash our skin every day we will soon start to smell. It is for the same reason that we should put on fresh clothes each day.

Soap removes dirt but it also removes natural oils leaving the skin dry. Some people rub **body lotion** on their bodies after washing to restore these oils.

The armpits are areas where we sweat most. **Deodorants** and **antiperspirants** counter the effect of sweat and keep our body smelling good throughout the day. Deodorants stop us smelling and antiperspirants help us to sweat less.

Our hair and the skin underneath which is called the scalp need to be regularly washed and combed. Some people use hair oils to prevent their hair becoming dry and brittle.

Finger nails and toe nails need regular trimming. Long nails may split and become infected.

Dirt should be regularly removed from the ends of nails and they should be washed with soap and a small nail brush to make sure they are really clean.

Teeth are an essential part of our digestive system and need regular cleaning. Apart from a toothbrush and toothpaste there are other products that help to remove food particles from between them.

FIG 6.9.1 Washing removes natural oils

FIG 6.9.2 Deodorants and antiperspirants keep us smelling good

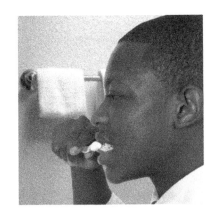

FIG 6.9.3 Teeth need regular cleaning

a)

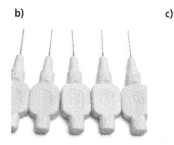
b)

c)

FIG 6.9.4 Products to help keep our teeth clean: **a)** dental tape **b)** interdental brushes **c)** mouthwash

Dental tape and **interdental brushes** help to remove small particles of food which your toothbrush might not be able to reach. **Mouthwash** rinses any remaining particles away and leaves your mouth and your breath feeling fresh.

Activity 6.9.1

Creating a kit for personal hygiene

Your teacher might put you into small groups for this activity.

Imagine that you are going to travel to another part of the country to visit a relative. They don't live near a store so you will need to take what you need for your own personal hygiene with you.

Make a list of the things that you will need and, alongside each one, say briefly how it is used.

Check your understanding

1. Briefly explain the use of each of the following products.

   a) Deodorant
   b) Dental tape
   c) Nail brush

2. Dirty dishes are washed in soapy water.

   Why might a person rub lotion into the skin on their hands after washing dishes?

Fun fact

The first deodorant was only marketed in 1888 and the first commercial antiperspirant in 1903. These are relatively new inventions.

Key terms

body lotion natural oils rubbed on the skin

deodorant prevents smell when a person sweats

antiperspirant reduces the amount a person sweats

dental tape tape that removes particles from between the teeth

interdental brush small brush for getting between the teeth

mouthwash cleans germs from the mouth leaving it feeling fresh

Review of Sexual maturity, reproduction and personal hygiene

- Growth is a characteristic of all living things.
- Human growth involves changes to the body other than increasing in size.
- A new-born baby is totally dependent on its mother and other adults.
- A child who is malnourished does not receive all of the nutrients in the quantities they need.
- Vaccination protects children from a range of diseases.
- Adolescence is a period in life between being a child and being an adult.
- Adolescents become sexually mature at puberty.
- Puberty is indicated by certain changes that take place to the body.
- During the years a woman can have children she has a regular menstrual cycle.
- The menstrual cycle is controlled by the action of the hormones oestrogen and progesterone.
- A woman's body can conceive long before she is mature enough to bring up a baby.
- During fertilisation a sperm combines with an egg to form a zygote.
- Fertilisation normally takes place in the oviduct.
- A zygote undergoes rapid cell division to become an embryo.
- An embryo is implanted in the thickened uterus wall.
- Personal hygiene is about looking after your body.
- Skin must be regularly washed.
- Natural oils can be replaced by body lotion.
- Hair must be regularly washed and nails regularly trimmed.
- Deodorants and antiperspirants help us to deal with the products of sweating.
- Teeth should be cleaned several times each day.
- Dental tape, interdental brushes and mouthwash can be used to keep the teeth clean.

Review questions on Sexual maturity, reproduction and personal hygiene

1. Name the parts A to E in Fig 6.RQ.1.

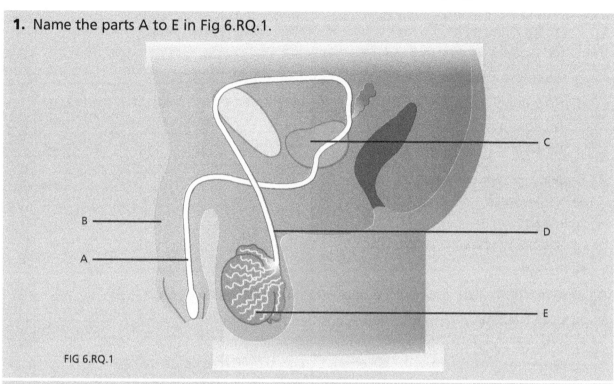

FIG 6.RQ.1

2. Name the parts A to E in Fig 6.RQ.2.

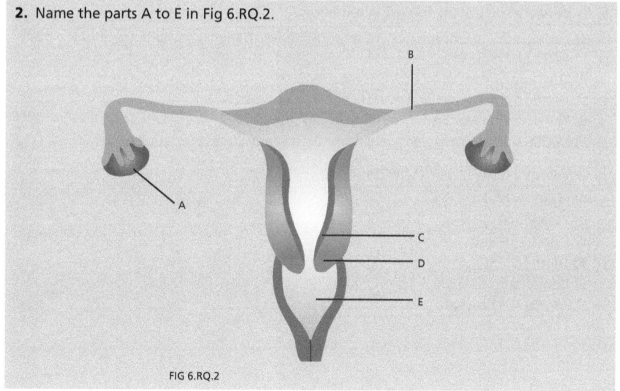

FIG 6.RQ.2

3. a) In which part of the body are the following formed?

  i) Sperm

  ii) Ova

b) What travels along:

  i) an oviduct?

  ii) a sperm duct?

c) What is the alternative name for a fallopian tube?

d) How many of the following are normally released by the body at any one time?

  i) Sperm

  ii) Ova

4. Explain the following terms.

a) Adolescence

b) Puberty

c) Secondary sexual characteristic

d) Zygote

5. a) Which hormones are involved in controlling the menstrual cycle?

b) State the role of each hormone identified in **a)**.

c) At what age does a girl usually start having periods?

d) What point of the menstrual cycle is usually considered to be day 1?

6. Copy and complete the following sentences.

a) _____ hormones are responsible for changes that take place to the body during adolescence.

b) The hormone released by the male sex organs is _____ .

c) During puberty the male voice becomes _____ .

d) The hormone released by the female sex organs is _____ .

e) During puberty the female starts to have regular _____ .

7. Explain the role of each of the following.

a) Mouthwash

b) Face lotion

c) Antiperspirant

d) Soap

e) Deodorant

f) Interdental brushes

8. The following diagram shows a part of the female reproductive system and some stages leading up to pregnancy.

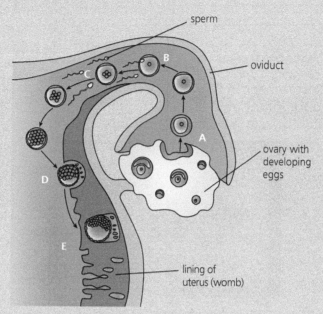

FIG 6.RQ.3

a) At what position marked A to E above:

   i)   is the egg released?

   ii) does fertilisation take place?

   iii) is the embryo formed?

b) What is happening at E?

Starter kits for mothers

Over 70 years ago the Government in Finland started to provide expectant mothers with a free pack in order to reduce infant mortality. At this time Finland was a poor country and a worrying number of children were dying during infancy.

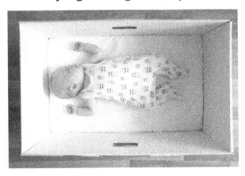

FIG 6.SIP.1 A bed for the new-born baby

The free pack is a starter kit for mothers and contains clothes, sheets and toys in a cardboard box. The cardboard box can even be used as a bed for the new-born baby.

The contents of the pack have been updated over the years as fashions change and different needs are recognised.

The Ministry of Health is giving consideration to operating a similar scheme in Jamaica. They have asked you to carry out research on their behalf to determine what should be included in such a pack.

1. You are going to work in groups of 3 or 4 to determine suitable contents for a new-born baby pack. The tasks are:

    • To review the needs of a new-born baby over the first months of life.
    • Use the Internet to find out what you can about the scheme that operates in Finland.
    • To draw up a provisional list of items to be included in the pack.
    • To carry out research by asking:
        ○ expectant mothers to comment on your list
        ○ experienced mothers to comment on your list.
    • To revise the list of items to be included in the pack on the basis of your research.
    • To compile a report for the Ministry giving your recommended list of items. This should include a short presentation in which you discuss your recommendations. You should illustrate your report by taking suitable pictures.

    a) Look back through the lessons in this unit which are about babies. Make sure you appreciate the needs of babies, and particularly very young babies.

    b) Find out what you can about the Finnish scheme using the Internet. A good place to start is www.bbc.co.uk/news/magazine-22751415. This will give you some background information about the scheme, why it was established and how it has changed over the years.

c) Draw up a list of things that you think would be useful for a mother to look after her new-born baby. In drawing up your list you should bear in mind such factors as:

- The climate in Jamaica is very different to that in Finland. This might affect your choice of clothing. For example, a new-born baby in Jamaica may not need woollen jumpers to keep warm.

- At the end of the day the box and its contents have to be paid for by the taxpayer. Whilst wanting to help the mother and baby as much as possible, there will be a limit on how much money is available.

- All of the items must fit into a box which can easily be carried by the mother.

d) Once you have drawn up your provisional list of contents for your box, and decided on the size of the box, you should ask the end-users for comments.

Show the list to one or more expectant mothers. They will obviously have given some thought as to what they will need then the baby is born so it will be useful for them, and for you, to compare lists. They might want things you have not included in your list while, at the same time, you might have things on your list that they didn't think about. Take some pictures of items that expectant mothers think they will need.

Show the list to one or more experienced mothers. They will have raised one or more children already and will be able to give you the benefit of their experience. You might find out from them that some things that were thought to be essential were never used while other things that weren't considered turned out to be really useful. Take some pictures of items that experienced mothers think are essential and others they think are of limited use.

e) Revise your list on the basis of what you learned from the expectant and experienced mothers. Bearing in mind the restraints on money and space, you might draw up a prioritised list in which you start with items you believe essential, such as disposable nappies for example, and then list items in order of decreasing importance.

You might draw a line across at some point on your list and decide to include those items above the line initially, and to include some items from below the line if there is sufficient money and space for them.

f) Prepare your report in which you should describe how you arrived at your recommended list of items. You should illustrate your report with photographs.

You might give a brief historical account of the Finnish scheme and explain how this has been modified to suit the requirements of modern-day Jamaica. You should explain how you have used the advice of expectant and experienced mothers in coming to your final recommendations.

Creating a 'personal hygiene' booklet

People who do not develop good habits of personal hygiene at an early age often suffer for it later in life. Treating the problems created by poor hygiene, such as skin conditions, dental decay and communicable diseases, is costly for regional health authorities and unpleasant for the individual.

The Ministry of Education have produced information booklets in the past, but they are concerned that these are not getting the message over to young people. The material is not taken as seriously as it should be.

In a fresh approach, the Ministry has decided to commission a 4-page booklet about personal hygiene targeted at children between 11 and 16 years, written and designed entirely by students of that age group. As you have recently been studying personal hygiene, you have been hired by the Ministry to produce this booklet.

1. You are going to work in groups of 3 or 4 to produce an interesting, attractive and factually correct 4-page booklet on personal hygiene that will appeal to your own age group. The tasks are:

   - To review the work on personal hygiene in the unit.
   - To research the range of personal hygiene products that are available and appropriate.
   - To identify several important areas that will be the foci of your booklet.
   - To write the factual content of the booklet.
   - To consider the design of the booklet.
   - To produce a 'mock up' of the booklet for comment by reviewers.
   - To produce a completed booklet.

   a) Look back through those parts of the unit that are concerned with personal hygiene. Make sure you are familiar with the different aspects that are covered.

   b) Spend some time looking at the personal hygiene products available in your local shops and markets.

   There are many to choose from. Avoid including long lists of products in your booklet as readers will not find this interesting, but if you are familiar with what products are available, it will help you compile the text.

FIG 6.SIP.2 Shops stock many personal hygiene products

   c) Decide which aspects of personal hygiene should be covered. You might decide, for example, that regular teeth cleaning is important because not doing this can have serious consequences. Conversely, although regular hair combing improves someone's appearance, it is less important for personal hygiene.

d) Once you have decided which aspects of personal hygiene are to feature in your booklet, the next task is to write the text. Here are some issues for you to consider:

- You have limited space, so comments need to be brief.
- You are writing for students of mixed ability, so use simple language that everyone can understand.
- Pictures can often be used in place of text. You could take some pictures of other students in your group carrying out activities such as cleaning their teeth.

e) The design of your booklet is a key feature of this product. You want to stimulate peoples' interest enough to want to pick up your booklet and read it. How are you going to achieve this?

f) Make a 'mock up' of your booklet by folding a sheet of letter-size paper in half. This will give you 4 pages. You can use white or coloured paper.

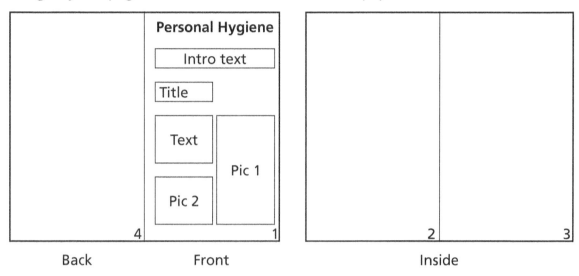

FIG 6.SIP.3 Layout of pages

You could start off by drawing a rough layout of how you envisage each page will look. This will give you an indication of how much room is available for text and what size your pictures should be. If you find there is too much text to fit the space available you will have to edit what you have written.

Print out your text in suitably sized text boxes and alter the sizes of pictures before printing to ensure they will fit in the picture boxes. Cut out the printed text and pictures and stick them into your booklet.

Don't forget to save your text files and pictures so that if you decide to make alterations you will not have to start again from the beginning.

g) Once you are happy with your first draft of the booklet, ask some reviewers to look through it and comment. You might ask them to look at particular aspects such as:

- Are there any typos (spelling mistakes, missing full stops, etc.)?
- Is the text as clear as it can be?
- Are the pictures meaningful?
- Overall, is the booklet attractive and interesting?

h) Once you get feedback from your reviewers, make whatever changes are necessary to the text and the pictures and produce a final version of your booklet.

Unit 7: Sexually transmitted infections and drug abuse

We are learning how to:
- describe some sexually transmitted infections
- understand drug abuse.

Sexually transmitted infections »»

Communicable infections are infections that can be passed on from one person to another. Sexually transmitted infections (STIs) are communicable infections of the reproductive system. They are passed on between people during sexual activity, and in particular during sexual intercourse. These infections are sometimes also called sexually transmitted diseases (STDs).

There are a number of STIs that it is possible for people to contract. Some of these, such as herpes, may cause considerable discomfort and inconvenience but they are not life-threatening. Other STIs, like syphilis, may result in severe illness and even death if they are not treated.

Condoms are a simple way of preventing the transfer of STIs, although their effectiveness depends on them being used correctly.

FIG 7.1.1 The transmission of STIs is preventable if people are prepared to take sensible precautions such as using condoms

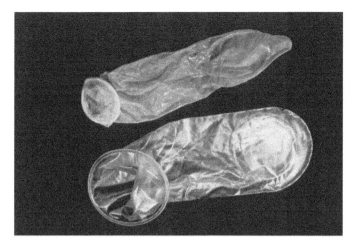

FIG 7.1.2 Condoms for men and women

Condoms are available for both men and women. The condom for men is slipped over the penis prior to sexual intercourse. The condom for women is inserted into the vagina.

Drugs 》》

A drug is a medicine or other substance which brings about physiological changes when we swallow it or otherwise introduce it into the body. Most people use drugs at times during their lives.

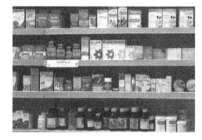

FIG 7.1.3 Drugs help us get over illness

Some people take drugs for a short period of time, such as to deal with the symptoms of a headache or a cold. Other people might need to takes drugs every day of their lives to help them cope with a disease or condition like diabetes.

A small minority of people take drugs for reasons other than illness. They claim that the drugs make them feel good and give them a 'high'. People may also abuse other substances, like inhaling solvents, for the same reason. This practice is known as drug abuse and is responsible for a number of deaths in Jamaica every year.

Something to think about 》》

There is a lot of misinformation about sexually transmitted infections and drug abuse. People who do not know very much try to hide their ignorance by making up stories or repeating stories they have heard.

There is a lot of reliable information on the Internet about sexually transmitted infections and drug abuse. There are also clinics where you can obtain information and talk to informed people.

Here is an example of a website you can try: kidshealth.org/teen

Herpes, chlamydia and human papilloma virus (HPV)

We are learning how to:
- describe the effects and prevention of sexually transmitted infections (STIs) and drug abuse
- describe herpes, chlamydia and human papilloma virus (HPV).

Herpes, chlamydia and human papilloma virus (HPV) 〉〉〉

Herpes simplex virus (HSV)

Herpes is caused by the herpes simplex virus (HSV).

Genital herpes is a common STI that results in blisters around the genitals of both men and women.

Herpes can be treated with the **antiviral drug** Aciclovir but the virus stays in the body and can recur.

Chlamydia

Chlamydia is caused by the bacterium *Chlamydia trachomatis*. Most people who have chlamydia do not experience any symptoms.

Chlamydia is treated with antibiotics.

Human papilloma virus (HPV)

The **human papilloma virus (HPV)** is a group of viruses that affect the skin and the moist linings of your body.

HPV is a highly contagious infection. There is no medical cure for the condition but the body's immune system is usually able to deal with it.

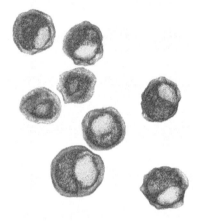

FIG 7.2.1 The bacterium *Chlamydia trachomatis*

Activity 7.2.1

Researching bacterial vaginosis

You will not require any equipment or materials for this activity, but you will need sources of reference material.

Here is what you should do:

Carry out research into the disease bacterial vaginosis. Use these questions to help you structure your answer.

1. What causes this disease?
2. What are the symptoms of this disease?

3. How are people infected?

4. What is the cure for this disease?

Check your understanding

1. The following passage is about a common STI called trichomoniasis. Read it carefully and answer the questions that follow.

FIG 7.2.2 *Trichomonas vaginalis* parasite

Trichomoniasis or 'trich' is caused by the protozoan parasite called Trichomonas vaginalis. *The parasite is passed from an infected person to an uninfected person during unprotected sex.*

Most women and men who have the parasite are not aware that they are infected.

When trichomoniasis does cause symptoms, they can range from mild irritation to severe inflammation. Men may feel itching or irritation inside the penis. There may also be burning after urination or ejaculation. Women may notice itching and discomfort during urination. They may also experience burning, redness or soreness of the genitals. The infection can be successfully treated with antibiotics.

   a) What causes the disease?

   b) Why might a person not realise that they are infected?

   c) How is the disease passed from person to person?

   d) What symptoms might an infected man experience?

   e) How is the disease treated?

Something to think about

Some STIs often produce few or no symptoms. People may have the disease for a long period of time and not be aware of it. During this time they might be infecting other people.

Key terms

herpes a sexually transmitted infection caused by the herpes simplex virus (HSV)

antiviral drug a drug that treats a virus

chlamydia one of the most common STIs, it is caused by the bacterium *Chlamydia trachomatis*

human papilloma virus (HPV) a group of viruses that affect the skin and the moist linings of the body

Gonorrhoea, syphilis and HIV

We are learning how to:

- describe the effects and prevention of sexually transmitted infections (STIs) and drug abuse
- describe gonorrhoea, syphilis and HIV
- present data using appropriate methods including tables and graphs
- draw conclusions from data.

Gonorrhoea, syphilis and HIV »

Gonorrhoea

Gonorrhoea is sometimes known simply as 'the clap'.

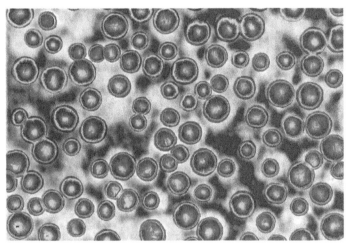

FIG 7.3.1 Gonorrhoea is caused by a bacterium called *Neisseria gonorrhoeae* or *gonococcus*

The typical symptoms of gonorrhoea include a thick green or yellow discharge from the vagina or penis, pain when urinating, and bleeding between periods in women.

The disease is usually treated with a single **antibiotic** injection and a single antibiotic tablet. If it is not treated there is a risk of serious complications, including infertility (inability to have children).

Syphilis

Syphilis is caused by a bacterium that is readily passed from an infected person during sexual activity.

There are three stages in the development of syphilis.

1. Primary syphilis: to begin with the disease is painless but there are highly infectious sores around the genitals. These last between two and six weeks before disappearing.

Fun fact

A number of famous people from history are thought to have died from syphilis, including Christopher Columbus, who was the first European to land in Jamaica in 1494.

FIG 7.3.2 Christopher Columbus

Scientists can never be certain without examining the person's body, but observations made by physicians, especially at the time a person was ill and died, sometimes correspond closely with the symptoms observed during the latter stages of untreated syphilis.

2. Secondary syphilis: the person remains infectious and develops symptoms such as skin rashes and a sore throat. These quickly disappear. After this the person will experience no other symptoms, perhaps for a number of years.

3. Tertiary syphilis: around one third of infected people who are not treated will, sooner or later, develop serious conditions, including heart disease, blindness, deafness, paralysis, insanity and eventual death.

Syphilis is treated using antibiotics, usually in the form of penicillin injections.

HIV and AIDS

AIDS stands for acquired immune deficiency syndrome. It is caused by a virus called the human immunodeficiency virus, or **HIV**.

The virus attacks the body's immune system, which is responsible for fighting diseases. This leaves the body unable to destroy the germs that cause other diseases.

You will learn a lot more about HIV and AIDS in future lessons.

Activity 7.3.1

Acquiring data about the instance of gonorrhoea, syphilis or HIV in Jamaica

You will not require any equipment or materials for this activity, but you will need sources of reference material.

Here is what you should do:

1. Choose one of the three STIs in this lesson as the focus of your research.

2. Obtain what data you can about the occurrence of this STI in Jamaica every year.

3. Represent the data you find in the form of a table and in a graphic form such as a bar chart or a graph.

Check your understanding

1. Answer the following questions about syphilis.

   a) What type of organism causes syphilis?

   b) During which stages is it infectious?

   c) Why might a person who has contracted syphilis incorrectly think they are no longer infected?

   d) What is the usual treatment for this disease?

Key terms

gonorrhoea sexually transmitted infection caused by bacterium called *Neisseria gonorrhoeae* or *gonococcus*

antibiotic medicine that works on bacterial infections

syphilis sexually transmitted disease caused by a bacterium that is readily passed from an infected person during sexual activity

AIDS stands for acquired immune deficiency syndrome

HIV human immunodeficiency virus, the virus that can cause AIDS

Interpreting data on STIs

We are learning how to:

- identify different ways in which data may be presented
- present and interpret data on STIs when presented in different ways.

Interpreting data

You may recall from Topic 1.12 that scientists display **data** in several different ways. They do this to show the patterns and relationships between data. It is much easier to interpret data in pictorial form than from lists and tables of numbers.

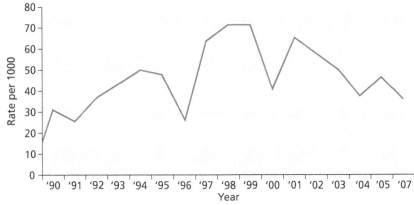

Source: National HIV/STI Control Programme Jamaica

FIG 7.4.1 **Prevalence** of HIV in women attending antenatal clinics in Jamaica 1990–2007

The graph shows the number of women per 1000 attending antenatal clinics who tested positive for HIV for the years 1990–2007 and who were not previously known to have the virus. The data could also have been presented as a table with columns for year and for rate per 1000, but it is much easier to see the pattern from the graph.

The graph shows that although there are ups and downs over the period, there is a general **trend** upwards to a maximum in 1998 and 1999 before the rate starts to fall again.

Sometimes related sets of data may be shown together. This makes the data more complex to interpret.

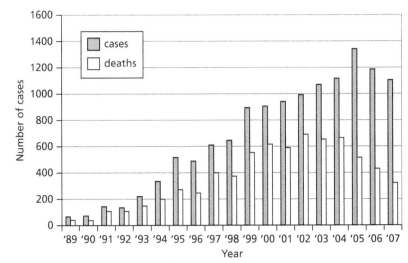

Source: National HIV/STI Control Programme Jamaica

FIG 7.4.2 Number of cases of AIDS and deaths resulting from AIDS in Jamaica 1982–2007

The bar chart in Fig 7.4.2 shows the number of cases of AIDS reported each year in Jamaica between 1982 and 2007 (red bars) and the number of deaths resulting from AIDS (black bars). From the heights of the bars we can identify trends in the number of cases, and in the number of deaths.

- The number of cases steadily rose up to 2004 and then remained around 1100.

- The number of deaths rose to a maximum around 2002–2004 and then started to fall.

As well as looking at trends in the two data sets, we can also make comparisons between them by looking at the heights of the bars for each year.

- The proportion of deaths to cases looks to be similar up to 2004 but then, although the number of cases remained more or less constant, the number of deaths fell. This might indicate the effects of new drugs used to treat AIDS.

Activity 7.4.1

Presenting data on AIDS

Table 7.4.1 shows the rate of AIDS in males per 100 000 persons in Jamaica between 1982 and 2007.

Age group / years	Rate of AIDS per 100 000 males
0–9	388
10–19	18
20–29	490
30–39	1283
40–49	1194
50–59	942
60+	308

TABLE 7.4.1

Present this data in a graphic form so that it is easy to see how the prevalence of AIDS differs in different age groups.

> **Fun fact**
>
> When using data obtained by somebody else it is usual to identify the source of the data. This acknowledges the work of others and also enables a reader to go back to the original source should they need to.

Key terms

data information

prevalence how many times something occurs

trend general direction

Check your understanding

1. The bar chart shows the prevalence of some STIs amongst three different groups of female sex workers.

   a) Which STI corresponds to the:

   i) blue bars?

   ii) brick-coloured bars?

   b) In which group of workers are STIs most prevalent?

   c) Which STIs were not found in clients?

   d) Overall, which two STIs were most common?

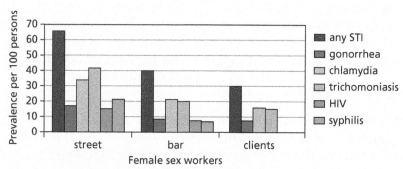

Source: Figueroa JP, Weir SS et al. Kingston PLACE RCT 2006 (modified)

FIG 7.4.3 Prevalence of STIs amongst different groups of female sex workers

217

Transmission and prevention of STIs

We are learning how to:

- describe the effects and prevention of sexually transmitted infections (STIs) and drug abuse
- appreciate the importance of responsible sexual behaviour
- evaluate the risks associated with irresponsible sexual behaviour.

Infections are described as sexually transmitted when they are passed from one person to another during sexual activity and, in particular, during sexual intercourse.

These infections spread most rapidly when people have sexual relationships with a number of different partners and when they have unprotected sex, i.e. sex without a condom.

FIG 7.5.1 Long-term partners

The simplest way to avoid STIs is not to have any sort of sexual relationship unless you are with a long-term partner and you both know that each other is not infected nor likely to be sexually active with others. The chances of becoming infected are much reduced if a person only has sex with a partner, in a stable relationship.

One of the most widely used methods of protection against STIs is **condoms**. They are available for both males and females. The condom provides a barrier between the **body fluids** of the participants during intercourse so tiny organisms like viruses and bacteria cannot pass from one person to the other.

FIG 7.5.2 Condoms provide some protection from STIs if they are used consistently and correctly

FIG 7.5.3 The waiting room in an STI clinic

Having unprotected sex once is enough for a disease to be passed on from one person to another. People who suspect they have been infected can go to a **clinic** where they will be examined and tests will be made. If treatment is necessary they can be given suitable drugs.

STI clinics can help to cure these infections but it is much better not to contract them in the first place. Many people end up suffering from an STI simply through lack of knowledge.

Avoidance is always better than cure.

Activity 7.5.1

Informing people about STIs

Here is what you need:

- Leaflets, pamphlets and other information available from STI and other clinics
- Large sheet of paper or thin cardboard.

Here is what you should do:

Use the printed material you obtain to make a poster informing people about one or more STIs and telling them how to avoid becoming infected.

Check your understanding

1. A young adult claims to have had unprotected sexual intercourse with several different partners and never caught any STIs. He thinks that all the warnings about STIs are unnecessary.

   What can you say that you think might persuade him to reconsider his attitude and become more socially responsible?

Key terms

condom a barrier between the fluids of the participants during sexual intercourse so organisms cannot pass from one person to another

body fluids fluids such as semen that are discharged by the body

clinic place where medical advice is given

avoidance stopping yourself from getting an infection

Drug use

We are learning how to:

- describe the effects and prevention of sexually transmitted infections (STIs) and drug abuse
- identify commonly used drugs
- distinguish between drug use, drug misuse and drug abuse.

The drugs people use 》》

We can classify the drugs that people commonly take into two groups:

- **Over-the-counter drugs** are drugs that adults can buy in shops without needing to consult their doctor or local clinic.

  People often buy mild painkillers like paracetamol over the counter to alleviate a headache.

- **Pharmacy drugs** are drugs that can only be bought with a prescription.

Drug misuse 》》

Although some drugs can be bought without a prescription you should not assume that they cannot harm you if taken incorrectly.

Drug misuse is when people take more than the stated dosage. Some people are foolish enough to believe that if they take two or three times the correct dosage then the drug will act more quickly or more effectively. This is not the case. If people exceed the stated dosage of a drug they risk damaging their bodies, or even death.

Alcoholic drinks contain a chemical called ethanol (alcohol) which acts as a drug when swallowed. We can also use the term alcohol abuse to describe the actions of people who drink rather more alcohol than is good for them.

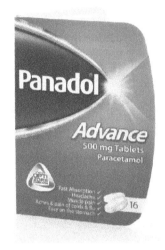

FIG 7.6.1 An over-the-counter drug

FIG 7.6.2 You need a prescription for a pharmacy drug

Activity 7.6.1

Over-the-counter-drugs

You will need to visit your local store or pharmacy to complete this activity.

Here is what you should do:

1. Make a list of the locally available drugs and medicines.

2. Organise them into groups according to how they are to be used. For example, you might have groups such as: 'Painkillers', 'Cold and flu treatment', 'Cough mixtures', etc.

3. Read the instructions for use on some of the packaging and confirm that the user is given sufficient advice for the correct use of the product.

Drug abuse »»

People who misuse drugs are abusing their bodies. While it can apply to both legal and illegal drugs, the term '**drug abuse**' is usually used to describe the habitual taking of illegal drugs. You will learn more about these drugs in the next lesson.

Check your understanding

1. Here are the recommended dosages for the drug paracetamol.

Children under 12 years old

10–15 mg per kg of body mass to be given every 4 hours

Maximum dosage 2000 mg per day

Children over 12 years and adults

1000 g every 8 hours

Maximum 3000 mg per day

   a) Suggest why the dosage is given per kilogram of body mass for children up to the age of 12 years but after this everyone gets the same amount.

   b) What is the maximum single dose to be given to a child of body mass 40 kg?

   c) How many times can this dose be safely given over a period of one day?

2. Karl has a cough. Should he take three times the prescribed dosage of cough mixture to relieve the cough because it is very severe? Explain your answer.

> **Fun fact**
>
> Painkillers are more correctly called analgesics. An analgesic drug is one that relieves pain.

Key terms

over-the-counter drug one that can be purchased without a prescription from a doctor or medical centre

pharmacy drug one that can only be purchased with a prescription

drug misuse taking drugs in an inappropriate way

drug abuse the habitual taking of illegal drugs

The ethics of drug research

We are learning how to:

- evaluate the importance of new drugs
- think about drug development in a balanced way.

New drugs >>>

Every year international **drug** companies spend many millions of dollars developing new drugs.

However, if you visit your local pharmacy you will see many shelves filled with different drugs. If we already have so many drugs, why do we need more?

There are two main reasons for drug development.

- New drugs may be more effective than existing drugs for treating certain conditions. For example, in the past aspirin was used to treat all pain, but now there is a range of painkillers to relieve different types of pain.

- New drugs may be needed to treat new conditions or diseases. During the world's first AIDS epidemic in the early 1980s, there were no effective drugs to combat this condition and death rates were high. Over 30 years later, the development of new drugs has greatly reduced death rates.

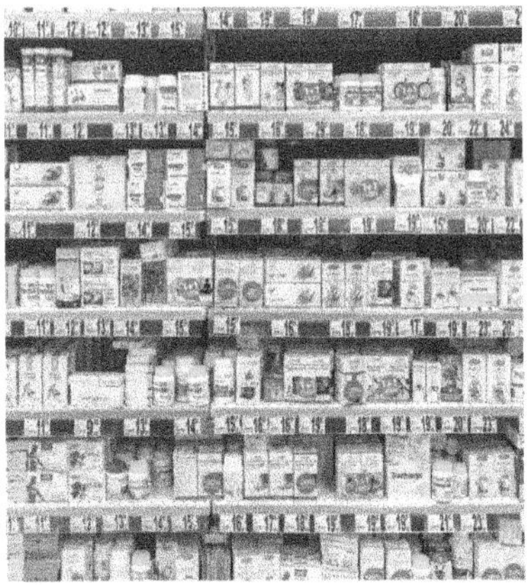

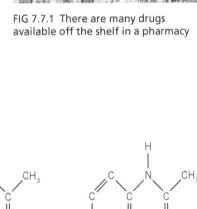

FIG 7.7.1 There are many drugs available off the shelf in a pharmacy

Drug development

When **synthesising** a new drug, a chemist is often guided by the structure of an existing drug. For example, there are similarities in the structures of the painkillers aspirin and paracetamol.

Newly synthesised drugs must be tested to ensure they are safe. These are the main stages in the development of a new drug:

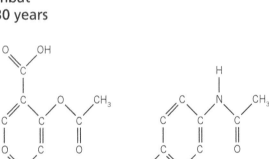

aspirin   paracetamol

FIG 7.7.2 Chemical structure of aspirin and paracetamol

synthesis → animal testing → clinical trials → general use

Once a drug is synthesised, it is tested on animals. It then undergoes clinical trials in which it is tested on humans. If the clinical trials prove successful, the drug may pass into general use.

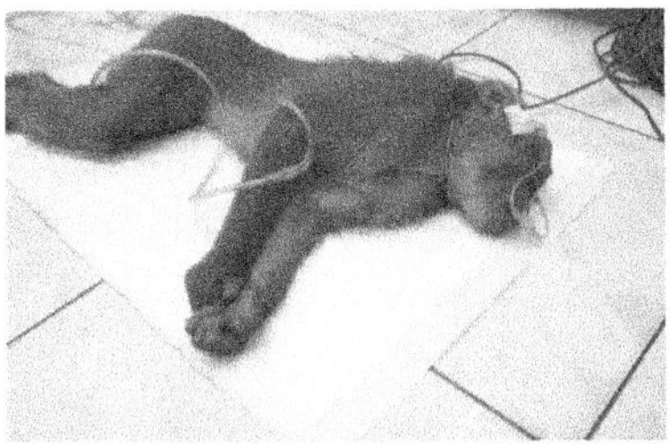

The ethics of animal testing

Many people have deeply held views about **animal testing**. Some people believe it is ethical or morally acceptable while others believe it is not.

FIG 7.7.3 Sometimes monkeys are used in animal testing, to find out if a new drug is **toxic**

Monkeys are sometimes used to test possible new drugs because their metabolism is closest to that of a human being. However, people argue that monkeys also share many of the feelings and emotions of humans.

Arguments against animal testing include:

- Many animals are killed or held in captivity.
- Animals are used to test substances that will ultimately not be useful.
- Animal testing is very expensive.
- Animals are never exactly the same as humans so results may not be valid.

Arguments in favour of animal testing include:

- Animal testing helps researchers to find new drugs and treatments.
- Animal testing improves human health and therefore life expectancy.
- Animal testing ensures that drugs are safe to use and will not have any unforeseen side effects.
- There are no alternative methods of testing which simulate human reaction in the same way.

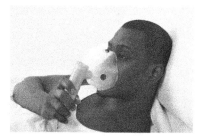

FIG 7.7.4 There are no known cures for some diseases

Fun fact

The term vivisection is sometimes used to describe animal testing.

Activity 7.7.1

Discussing the ethics of using animals in medical research

Your teacher will organise you into small groups for this activity. Here is what you should do:

1. Discuss how you feel about animal testing.

2. Try to get a balanced view between what some regard as cruelty to animals, and the needs of society. You need to address questions like 'Does saving human life justify the use of animals in this way?'

3. Decide whether your group is going to be in favour of or against animal testing. Write down some arguments which support the view of the group.

Key terms

drug substance that has a physiological effect when introduced to the body

synthesis production of compounds by chemical reactions

toxic poisonous

animal testing use of animals to test if substances are beneficial or harmful to humans

Check your understanding

1. Some cosmetics companies use animals to test if their products irritate the skin and eyes. Other companies clearly state that animal testing has not been used.

   a) Why might a person view the use of animals to test cosmetics as different to using animals to test drugs?

   b) If a cosmetic is advertised as having been tested on animals, do you think it will sell better or less well? Explain your answer.

Drug abuse

We are learning how to:

- describe the effects and prevention of sexually transmitted infections (STIs) and drug abuse
- describe the dangers of drug misuse and abuse
- explain drug addiction and rehabilitation.

Illegal drugs >>>

Drug abuse can involve taking drugs which have been declared **illegal** because they are potentially harmful to people and often result in **addiction**.

There are many illegal drugs in use around the world which users swallow, smoke or inject according to what it is. Some of the commonest are cannabis (**ganja**), heroin, cocaine, methamphetamines and ecstasy.

The use of ganja has links to the Rastafarian faith and in 2015 this drug was decriminalised. In Jamaica today, the cultivation of up to five plants is permitted and possession of up to 50 g is considered a petty offence which will not result in a criminal record. Large amounts of this, however, remain illegal.

The trade in and use of illegal drugs are criminal offences. Use of these drugs is prohibited because they are a danger both to individuals and to society.

Illegal drugs are dangerous to the individual because:

- they damage the body causing both short-term and long-term health problems
- they impair a person's judgement so they become a danger to themselves and to others
- there is no control over the manufacturing process so drugs could contain other harmful substances
- the effects of a drug might be different from person to person
- they are addictive so once a person becomes a regular user they cannot stop using the drug
- a user may accidently or deliberately take an overdose, which can result in death.

FIG 7.8.1 Possession of small amounts of ganja is overlooked in Jamaica

FIG 7.8.2 There are many types of illegal drugs and they affect both individuals and society as a whole

Illegal drugs are dangerous to society because:

- they entice users to break the law when they buy the drugs

- users who are addicts might commit crimes to obtain the money for drugs

- they take large amounts of money out of the economy on which no taxes are paid

- they generate large amounts of money that can be used for other illegal activities

- they create a gang culture in which different groups fight to control the drug trade

- innocent people may become caught up in the gang warfare.

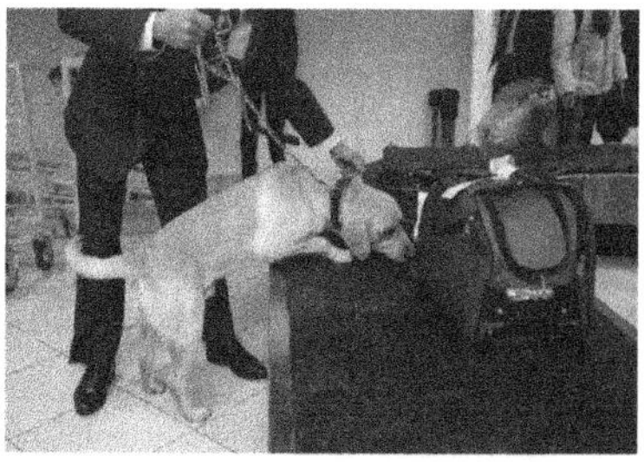

FIG 7.8.3 Law enforcement agencies confiscate and destroy illegal drugs

Activity 7.8.1

Role play on drug abuse

Your teacher will organise the class into small groups for this activity.

Here is what you should do:

1. Choose one of the group to be a person who is just going to try out a drug once, just to see what it is like.

2. The rest of the group are his/her friends/relatives and it is their job to dissuade him/her from this foolish plan.

3. Write a short script around this scenario which you will then be invited to perform in front of the class.

Drug addicts who really want to change their life style can find help to escape their addiction to drugs at a **rehabilitation centre**. At such a centre their reliance on the drug is slowly reduced under medical supervision. At the end of their treatment they are said to be 'clean' and can go back to leading a normal drug-free life.

Fun fact

In some countries cannabis is used as a medicine. Amongst other things it is thought to relieve chronic pain and muscle spasms.

Key terms

illegal against the law

addiction a physical dependence

ganja another name for cannabis

rehabilitation centre place where an addict goes to overcome their addiction to a drug

Check your understanding

1. Name three illegal drugs.

2. a) Give one argument against the decriminalisation of ganja in small amounts.

   b) Give one argument for the decriminalisation of ganja in small amounts.

3. Give three ways in which the traffic in illegal drugs is detrimental to society.

Review of Sexually transmitted infections and drug abuse

- Communicable diseases of the reproductive system are often called sexually transmitted diseases (STDs) or sexually transmitted infections (STIs). These diseases are described as sexually transmitted because they may be transmitted from one person to another during sexual activity, and in particular sexual intercourse.

- Herpes, chlamydia and the human papilloma virus (HPV) are examples of common STIs.

Disease	Cause	Symptoms/Effects	Treatment
Genital herpes	Virus	Skin damage or blisters round the genitals	Antiviral drug Aciclovir
Chlamydia	Bacterium	Number of symptoms, though some sufferers may be unaware that they have it	Antibiotics
Human papilloma virus	Group of viruses	Affects moist membranes of the body – may lead to genital warts and cervical cancer in women if left untreated	No medical treatment but body's immune system is usually able to deal with it
Gonorrhoea	Bacterium	Thick green or yellow discharge from vagina or penis	Antibiotics
Syphilis	Bacterium	Sores on the genitals – if not treated serious conditions such as heart disease, blindness, deafness, paralysis and insanity may develop	Antibiotics, e.g. penicillin

TABLE 7.R.1

- Data on STIs may be presented in different ways.

- The simplest way to avoid STIs is not to have any sort of sexual relationship unless you are with a long-term partner and you each know that the other is not infected or likely to be sexually active with other people.

- Condoms provide some protection against STIs by preventing the transfer of body fluids.

- Many people end up with an STI because they lack knowledge. The number of cases of STIs each year can be reduced by educating people. It is better to avoid becoming infected even if there is a simple cure.

- Drugs are chemicals that bring about physiological changes when introduced to the body.

- Over-the-counter drugs can be bought without a prescription from a doctor.

- Pharmacy drugs can only be bought with a prescription from a doctor.

- People use over-the-counter drugs and prescription drugs when they are unwell.

- All drugs are potentially harmful if the instructions for use are not followed.

- There are several stages in the development of a new drug.

- Research into new drugs involves ethical issues like the use of animals.

- Drug misuse involves taking drugs in an inappropriate way. This includes drugs like ethanol, which is found in alcoholic drinks.

- Drug abuse is about taking illegal drugs, but can also be about taking too many legal drugs.

- Illegal drugs include ganja, heroin and cocaine.

- People are permitted to have small quantities of ganja in Jamaica.

- Drug abuse is bad for the individual for a number of reasons.

- Some drugs are addictive and people become physically dependent on them.

- Drug addicts can be cured at a drug rehabilitation centre.

- Drug abuse creates many problems for society.

Review questions on Sexually transmitted infections and drug abuse

1. **a)** What does the abbreviation STI stand for?
    b) State whether each of the following is caused by a bacterium or a virus.
      i) HIV
      ii) Syphilis
      iii) Chlamydia
      iv) Genital herpes

2. Which of the following statements are true and which are false? Write T or F for each one.

    a) The body will cure itself of STIs eventually without the need for medical help.
    b) Condoms are available for both men and women.
    c) Taking twice the dosage of a painkiller will make it twice as effective.
    d) People in Jamaica can grow as much ganja as they like within the law.
    e) A drug suitable to treat one STI will not work on other STIs.
    f) Some drugs are addictive and the user becomes physically dependent on them.

3. The following table shows the number of deaths due to AIDS per 100 000 people in Jamaica over the period 2005–2015.

Year	Number of people who died of AIDS per 100 000 people in Jamaica
2005	19.4
2006	16.3
2007	12.0
2008	15.0
2009	14.1
2010	12.4
2011	14.6
2012	9.6
2013	11.0
2014	8.2
2015	9.4

TABLE 7.RQ.1   (Source: Ministry of Health Jamaica)

    a) Draw a bar graph to represent this information.
    b) Does the data suggest that the fight against AIDS over this period was successful? Explain your answer.
    c) Assuming the trend continues in the same way, predict a likely value for 2018.

4. **a)** What causes gonorrhoea?

   b) What are the symptoms of this disease?

   c) What is the treatment for this disease?

   d) What complications might occur if gonorrhoea is left untreated?

5. Ibuprofen is an over-the-counter drug that people commonly buy in Jamaica.

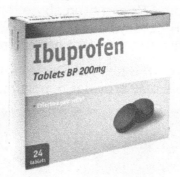

FIG 7.RQ.1

   a) State three reasons why a person might take this drug.

   b) What is the adult dosage for this drug?

   c) What is the maximum number of tablets an adult can take over 24 hours?

   d) What advice is given about giving this drug to children under 12 year of age?

6. The following is an extract from a report by the International Narcotics Control Board (INCB) in 2015.

   Jamaica remains the largest illicit producer and exporter of cannabis herb in Central America and the Caribbean and accounts for approximately one third of cannabis herb produced.

   Jamaica has also become a hub for the trafficking of cocaine, owing to the displacement of trafficking routes as a result of the strengthening of drug trafficking countermeasures in Latin America. Jamaican criminal groups are using the elaborate networks originally established to traffic cannabis and cocaine as well.

   In Jamaica, drug trafficking takes place at airports through drug couriers, baggage and air freight and at seaports via containers, cargo vessels, underwater canisters attached to ship hulls, shipping vessels and speedboats. Illicit drugs are traded for money, guns and other goods, and much of the proceeds are used to foster criminal activities.

   a) i) For which drug is Jamaica the largest producer and exporter?

   ii) Approximately what percentage of the overall production comes from Jamaica?

   b) i) For which other drug has Jamaica become the centre of trafficking?

   ii) Why has this shifted from other parts of Latin America?

   c) State two places where drugs are trafficked and give one example of how this is done for each place.

   d) What happens to the money obtained by trafficking in illegal drugs?

Making a statement through role play

It is difficult to inform people about STIs and drug abuse if they find booklets and pamphlets boring and refuse to read them. They may have reading difficulties and cannot understand what the text means, or maybe they don't think it has any relevance to them and their lives.

For these people, a different approach is needed. One alternative is role play. Lots of people enjoy watching drama. Role play provides opportunities to pass on information in an informal way. If the role play is well planned and well written, the audience will come away with important messages about STIs and drug abuse that might stop them from doing foolish things in the future.

A touring theatre company who put on plays at schools like yours has approached you with a request that you use your knowledge to write a short play which will inform young people about the problems associated with STIs or drug abuse.

FIG 7.SIP.1 Role play provides a stimulating way to learn

1. You are going to work in groups of 3 or 4 to write a short play. The tasks are:

   - To review the problems associated with STIs and drug abuse which are described in this unit.
   - To look at examples of short plays.
   - To decide the age group for which you will write your play.
   - To decide how long your play will be.
   - To write the script for your play.
   - To determine what props, if any, you will need.
   - To trial your play to a small audience of advisers.
   - To modify your play on the basis of feedback from the trial.
   - To prepare a report for the theatre company on the work you have carried out. This should include a demonstration of the final version of your play.

   a) Look back through the unit and identify a key message that you would like to deliver about STIs or drug abuse.

b) Look for examples of short plays. The drama teacher might be able to help, or you can look on the Internet.

Your play could have a variety of endings, such as happy, sad, justice is done, good prevails, etc. Which will give the greatest impact? How will you ensure your play makes an impact and doesn't put the audience to sleep? The messages that make the most impact are those that are simple to understand.

FIG 7.SIP.2 Is this message clear?

c) Decide on a target age group for your play. For example, you might think that 11–16 years is a suitable age range. Some role play will be understood by a wide age range, whereas some themes are more suitable for older students.

d) Decide how long your play will be. It needs to be long enough to get your message over to the audience but not so long that people lose interest or become side-tracked by secondary plots. This sort of role play often lasts for 3–5 minutes.

e) The next task is to write your script. Don't worry about props to start with. You need to find a story line that works and to create the words. Try to give everyone in the group a part.

For example, how about:

- A student becomes involved in buying drugs.
- His younger brother/sister accidentaly finds drugs in his bedroom.
- He accuses her of snooping.
- She denies this and persuades him to...

Once you have a story line, identify who will play each role and write the lines they will say.

Once the play is written think about props. It won't be possible to create large items such as backdrops, but if some small props, such as a syringe, would enhance your play, then see if you can find or borrow them.

f) Identify a group of people who you will ask to watch your play. These people should not have been involved in writing the play. Ask them to take some photographs of your performance, which can be used to illustrate your final report.

Ask each person for their opinion after they have watched the play. Don't be defensive about criticism as this is the purpose of trialling the play. If the message isn't clear or the lines don't make sense, now is the time to sort them out.

g) Revise your play, incorporating any changes needed as a result of the trial. Your play will form part of your final report so it is worth spending time getting it right.

h) Prepare a final report which gives details such as:

- What your play is about and how long it lasts.
- What age group it is designed for.
- What message it seeks to deliver about STIs or drug abuse.

Use the photographs of your group's performance to illustrate your report.

Unit 8: Climate change

We are learning how to:
- describe some changes to climate
- explain some observed changes.

Environmental impact of human activity ▶▶

There are many ways in which humans bring about changes to the environment. The effects of some human activity go beyond altering local habitats and ecosystems. They bring about global changes which affect everyone on the Earth no matter where in the world they live.

In this unit you are going to consider how the Earth's climate is changing, what is the likely cause, and what impact this will have, not just on humans, but on biodiversity.

Global warming

As a result of increases in atmospheric carbon dioxide and other gases, the surface temperature of the Earth is increasing by a small but significant amount each year. The increase in temperature is called global warming.

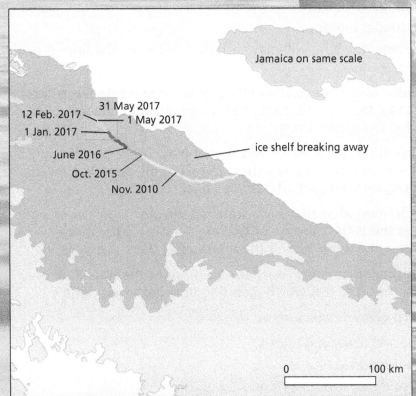

FIG 8.1.1 Larsen C Ice Shelf in Antarctica released the outlined iceberg in July 2017

As a result of global warming the Earth's poles are getting warmer. A crack has progressively grown across the Larsen C Ice Shelf. Eventually an iceberg half the size of Jamaica will float away.

Greenhouse effect

Global warming is thought to be the result of the greenhouse effect. Too much heat is being trapped on Earth by the atmosphere and not enough is escaping into space. This is similar to what happens in a greenhouse.

Climate change

As a result of global warming the climate in different parts of the world is changing. Some areas of the world are receiving higher than normal rainfall while others are suffering droughts. Average winter temperatures in many places have risen, which has led to changes in the growing cycles of plants and animals.

FIG 8.1.2 Bush fire in East Rural St Andrew

FIG 8.1.3 Bog Walk 1865

Deforestation

Large areas of forest are being cut down every year to provide wood and farmland.

Forests provide lots of different habitats. Once these habitats are lost the animals that occupy them are also lost.

Forests have an important role in extracting carbon dioxide from the atmosphere. Smaller areas of forest mean less carbon dioxide is removed, which enhances the greenhouse effect.

Between 1900 and 2000 Jamaica lost an average of 400 hectares of forest every year.

Something to think about

Global warming has brought about the realisation that people cannot live in one part of the Earth in isolation. Some human activities can affect everyone and therefore a global strategy is needed to protect our planet.

Global warming

We are learning how to:

- explain climate change in terms of human activity
- consider the possible effects of global warming.

Global warming >>>

The average surface temperature of the Earth has increased by about 0.8 °C over the last 100 years. Much of this increase has occurred in the last 30 years. This suggests that the trend in rising temperature is increasing.

This may not sound like much of an increase, but even this small rise has caused major changes to the climate in different parts of the world. This is known as **global warming**.

FIG 8.2.1 The North Pole is at the extreme top of the Earth and is a very cold place, even in the summer

The average winter temperature at the North Pole is around −34 °C while in the summer the average is around 0 °C. Scientists who study the North Pole have been aware for some time that average temperatures are increasing. The northern polar ice cap is slowly getting smaller and the ice is getting thinner as more ice turns to water.

In some parts of the world the summers are getting much hotter and drier. Warmer weather has led to water shortages and **droughts** in certain parts of the world.

Water is being used up more quickly than it can be replaced by nature.

FIG 8.2.2 The level of water in reservoirs is lower than it has been in the past because of increased evaporation and lack of rain

In areas of the world where there are large forests, the vegetation is much drier than in the past due to higher temperatures and lack of rain.

FIG 8.2.3 Dry vegetation catches fire very easily and forest fires have destroyed huge areas of forest

Activity 8.2.1

The effects of drought

You should work in a small group for this activity.

Here is what you should do:

1. Discuss what effects a drought is likely to have on people and on wildlife.

2. Imagine there is a drought in your area now. Discuss what steps you could take to reduce the use of water.

Check your understanding

1. Copy and complete the following sentences by writing either 'increasing' or 'decreasing'.

a) The surface temperature of the Earth has been _____ over the last 100 years.

b) The thickness of the ice at the North Pole is _____ .

c) In some parts of the world summers are getting hotter and drier. The effects of this are that:

i) water levels in lakes and reservoirs are _____ .

ii) the risk of forest fires is _____ .

iii) the number of wild animals and farm animals dying from lack of food and water is _____ .

> **Fun fact**
>
> There is land beneath the ice at the South Pole, but the North Pole consists entirely of ice. If the northern polar ice cap were to completely melt it would be possible to sail over the North Pole.

Key terms

global warming rise in average surface temperature of the Earth

droughts lack of rain

The greenhouse effect

We are learning how to:

- explain climate change in terms of human activity
- explain the greenhouse effect.

The greenhouse effect 〉〉

Greenhouse gases are gases that can absorb and emit heat radiation. The main greenhouse gases in the **atmosphere**, as far as global warming is concerned, are water vapour, carbon dioxide, methane and ozone.

Many scientists believe that global warming is caused by increasing concentrations of greenhouse gases in the atmosphere. This increase is the result of human activities.

The increase in concentrations of greenhouse gases in the atmosphere is commonly called the **greenhouse effect** because these gases have a similar effect to the glass roof and sides of a greenhouse. Heat becomes trapped in the greenhouse causing the temperature to rise.

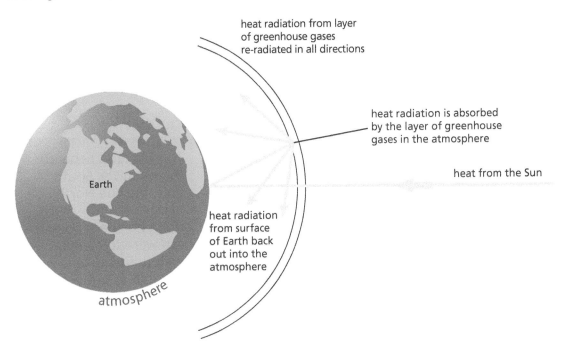

FIG 8.3.1 The greenhouse effect

The greenhouse effect should more correctly be called the enhanced greenhouse effect. The greenhouse effect has existed as long as the Earth has had an atmosphere. Without the greenhouse effect, Earth would never have become warm enough to support life as we know it.

The problem is that concentrations of greenhouse gases have increased significantly over the past 200 years. Now, too much heat is being trapped on the Earth by the greenhouse gases and not enough is escaping into space.

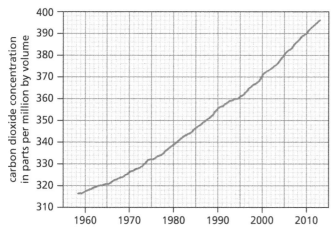

FIG 8.3.2 Global warming coincides with a slow but steady rise in the concentration of carbon dioxide in the atmosphere, which provides evidence of the link between global warming and the effect of increasing concentrations of greenhouse gases

Activity 8.3.1

Tackling global warming

You should work in a small group for this activity.

Here is what you need:

- Plain paper
- Coloured pencils or paints.

Here is what you should do:

Global warming is a global problem but that does not mean individuals cannot take action to reduce it.

1. What action can you take as an individual to reduce global warming?

2. Design a leaflet informing people about the problems of global warming and what they can do to help reduce it.

Check your understanding

1. State whether each of the following increase or decrease the concentration of carbon dioxide in the air.

   a) Burning fossil fuels like coal and natural gas
   b) Photosynthesis
   c) Road vehicles that run on petrol and diesel
   d) Deforestation

Fun fact

Concentration is sometimes expressed as the number of parts per million. For example, the concentration of carbon dioxide is 396 parts per million.

This means that for every 1 000 000 particles in the air, 396 of them are carbon dioxide. This can also be expressed this as a percentage:

$$\frac{396}{1\ 000\ 000} = \frac{0.0396}{100}$$

$$= 0.0396\%$$

Key terms

greenhouse gases gases that can absorb and emit heat radiation

atmosphere layer of gases above the surface of the Earth

greenhouse effect the increase in concentrations of greenhouse gases in the atmosphere

Greenhouse gases

We are learning how to:

- explain climate change in terms of human activity
- assess the impact of different greenhouse gases.

Greenhouse gases

You are already aware that a layer of greenhouse gases exists in the atmosphere. This layer allows heat radiation from the Sun to heat the Earth, but traps some of the heat which passes from the Earth back into space.

The greenhouse effect in itself is not a bad thing. By trapping heat it has allowed the Earth to warm up to the point where life can exist on it. The problem now is that this layer has become too efficient at trapping heat and the Earth is heating up too much.

In the previous lesson you learned about the link between global warming and the rise in the concentration of carbon dioxide in the atmosphere. Other atmospheric gases which make a significant contribution to the greenhouse effect include methane, nitrous oxide, ozone and chlorofluorocarbons (CFCs).

Greenhouse gas	Relative effect per molecule	Concentration / parts per million by volume
Carbon dioxide	1	387
Methane	30	1.7
Nitrous oxide	160	0.31
Ozone (lower atmosphere)	2 000	0.06
CFC 11	21 000	0.00026
CFC 12	25 000	0.00044

TABLE 8.4.1 Greenhouse gases   (Source: UK Department of the Environment)

Since the start of the industrial revolution, concentrations of greenhouse gases like water vapour, carbon dioxide and methane have been rising. CFCs, which are responsible for damage to the ozone layer, are also greenhouse gases.

Notice that there are two factors which determine the contribution of a greenhouse gas: the relative effect it has in trapping heat, or the **greenhouse index**; and the concentration or **abundance** of the gas found in the atmosphere.

> **Fun fact**
>
> Many people make good use of waste plant material and help the environment by having a compost bin or a compost heap. The waste plant material decays and releases nutrients that are used to enrich soil. However, methane is also a product of natural decay and is produced during composting so they are also contributing to the rising level of atmospheric methane.

Finding which gases make the largest contribution to the greenhouse effect

8.4

Table 8.4.2 shows the greenhouse index of some gases in the upper atmosphere compared to carbon dioxide. On this scale carbon dioxide = 1.0.

The contribution of each gas to the greenhouse effect depends both on its index and its abundance in the upper atmosphere.

1. Copy and complete the table.

2. State which two gases make the greatest contribution to the greenhouse effect.

3. State which two gases make the least contribution to the greenhouse effect.

Gas	Greenhouse index	% abundance in upper atmosphere	Index × abundance
Carbon dioxide	1	0.04	0.04
CFCs	23 000	0.00000004	
Methane	30	0.00017	
Nitrous oxide	160	0.00003	
Oxygen	0	21.0	0
Ozone	2 000	0.000004	
Water vapour	0.1	1.0	

TABLE 8.4.2

It is not necessarily the gases which have the highest ability to trap heat which are the biggest contributors to the greenhouse effect since they may only be present in low concentrations. CFCs are 23 000 times more effective than carbon dioxide but they are present in such low concentrations that their overall contribution is very small.

Conversely, a gas with a low ability to trap heat might be an important greenhouse gas because it is present in high concentrations. Nitrous oxide is much less effective at trapping heat than CFCs but its concentration in the atmosphere is around 1000 times higher.

Key terms

greenhouse index relative contribution made to the greenhouse effect

abundance amount or concentration of a substance

1. Table 8.4.3 shows the average concentration of methane in parts per billion (ppb) for June for the period between 2006 and 2015.

   a) Plot a graph of concentration of methane on the y-axis and the year on the x-axis.

   b) What is the general trend in concentration of methane over this period?

   c) Does this data provide proof that methane is a greenhouse gas? Explain your answer.

Year	Concentration of methane in ppb
2007	1781
2009	1806
2011	1810
2013	1815
2015	1845

TABLE 8.4.3   (Data from Muna Loa Observatory)

Impact of climate change

We are learning how to:

- explain climate change in terms of human activity
- consider the impact of climate change.

Climate and weather ⟫

Climate and **weather** are different things and should not be confused. The weather is about what is happening now at a particular place. We might describe the weather in Kingston today as wet or dry, warm or cold, windy or calm.

The climate is concerned with weather conditions over a long period of time in a large area. It is more general and gives a more overall picture. When we say that the climate in Jamaica is generally warm and sunny this doesn't refer to a particular day but to most days.

FIG 8.5.1 Sunny Kingston

Climate change

Although we might say that the climate in an area is warm and dry in the summer and cold and wet in the winter this doesn't mean that every summer day is equally warm and dry and every winter day is equally cold and wet.

You might have heard people say things like 'last summer was hotter than usual' or 'there wasn't much rain last winter'. Within any area there are natural fluctuations in the climate. When global warming and climate change was first suggested not all scientists were convinced that the greenhouse effect was to blame. They argued that what countries were experiencing was simply natural variation.

The situation has now become clearer and most scientists accept that global warming is responsible for unseasonal weather conditions that have affected large areas of the world over the past decade.

In 2012, a study of the climate in Jamaica found that the frequency of warm days, warm nights and extreme high temperatures had increased while there were fewer cool days, cool nights and extreme low temperatures.

In other parts of the world, the climate change has been far more dramatic. Here is an example.

Glaciers are found near the tops of high mountains or at the Earth's poles where it is very cold. They are sometimes described as rivers of ice. Snow accumulates over time, continually forming huge sheets of ice that flow very slowly down the mountain. The top of the glacier is continually replenished while the bottom will eventually melt as it warms up.

FIG 8.5.2 Melting glacier

Global warming has accelerated the rate with which glaciers are melting all around the world. At first glance this might seems a good thing since it releases lots of fresh water into the environment. However, this also causes problems.

- Valleys into which glaciers flow have now become lakes and local environments have been lost.

- Where the water flows out to sea, the salinity decreases, which affects marine organisms.

- The level of the sea is rising, which affects low-lying countries like the Maldives in the Indian Ocean.

Activity 8.5.1

How might climate change in Jamaica affect my life?

Your teacher will organise the class into small groups for this activity.

Within your group discuss how you think your life might be affected if there are long-term climate changes in the Caribbean. You might consider:

Advantages like: more sunlight = more electricity from solar power plants

Disadvantages like: less rain = more expensive vegetables and fruits

Check your understanding

1. The following pictures show the Muir Glacier in Alaska, USA in two different years.

FIG 8.5.3

a) How many years passed between the two pictures?

b) What is the main change that can be seen in the pictures?

c) Suggest a reason for this change.

> **Fun fact**
>
> On the planet Venus, the greenhouse effect traps so much heat energy that the average surface temperature is 462 °C.

Key terms

climate conditions like average temperature, average rainfall over an area over a long period of time

weather local conditions on a day like temperature, amount of rainfall, wind direction

glacier extremely large mass of ice which moves very slowly, often down a mountain valley

Reducing climate change

We are learning how to:

- describe ways of reducing greenhouse gases
- describe other initiatives to mitigate climate change.

Greenhouse gases and climate change ›››

Since there are clear links between greenhouse gases, global warming and climate change, any initiative to reduce climate change must involve a reduction in the levels of greenhouse gases. The contribution of an individual gas depends on its greenhouse index value and its concentration in the upper atmosphere.

Although the greenhouse index of carbon dioxide is small, it is present in a relatively high concentration in the upper atmosphere. The gradual increase in atmospheric carbon dioxide over the past century has been closely associated with global warming, so we will focus on initiatives for reducing the concentration of this gas.

FIG 8.6.1 Plants and the oceans are natural carbon dioxide sinks

Carbon dioxide sinks

A **carbon dioxide sink** is a carbon reservoir that increases in size. The main natural sinks are:

- plants and other organisms
- the oceans into which carbon dioxide dissolves

Any initiatives involving **reforestation** in the Caribbean, will increase the capacity of natural sinks.

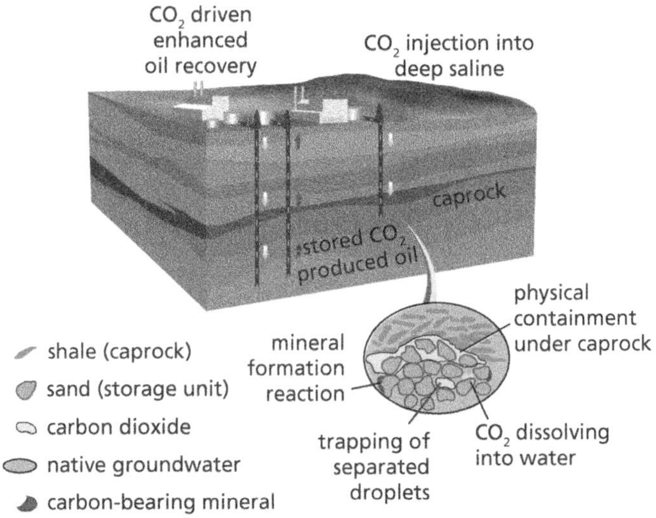

There are now plans to remove carbon dioxide from the atmosphere and store it in traps under suitable layers of rock.

CO_2 driven enhanced oil recovery

CO_2 injection into deep saline

caprock

stored CO_2
produced oil

shale (caprock)
sand (storage unit)
carbon dioxide
native groundwater
carbon-bearing mineral

mineral formation reaction

trapping of separated droplets

physical containment under caprock

CO_2 dissolving into water

FIG 8.6.2 Carbon dioxide capture

Electrically powered motor vehicles

The biggest single source of atmospheric carbon dioxide is the combustion of carbon-based fuels such as coal, natural gas and those made from crude oil, such as diesel and petrol. Any initiative to reduce the numbers of vehicles powered by diesel and petrol engines will be hugely beneficial.

Scientists and technologists have experimented with electrically powered cars for many years. They don't burn a fuel and therefore don't produce carbon dioxide.

FIG 8.6.3 Electrically powered car

Recent developments in **battery** technology mean that electric cars are now possible and the numbers in use will rapidly increase over the next decade. Scientists are also working on electrically powered trucks.

Alternative energy sources

Fossil fuels are also burnt on a large scale in power stations to generate electricity. The Taichung Power Plant in Taiwan is the world's biggest coal-fired power station, and also the world's biggest producer of carbon dioxide. It releases 40 million tonnes of carbon dioxide into the air every year.

Any initiative that promotes the use of alternative energy sources will reduce atmospheric carbon dioxide.

FIG 8.6.4 Taichung Power Plant in Taiwan

Activity 8.6.1

My carbon footprint

You will not need any equipment or materials for this activity. You should work in groups of 3 or 4.

1. Carry out some research to find the meaning of the term 'carbon footprint'.

2. Make a list of the ways in which you, and your family, contribute to your own personal carbon footprint.

3. Discuss in your group how, as an individual, you can reduce your carbon footprint.

Fun fact

Biofuels, like biodiesel, are sometimes described as 'carbon neutral' because the amount of carbon dioxide released when they are burnt is equal to the amount of carbon dioxide absorbed by the plants when they are growing.

Check your understanding

1. Read the following extract from an article in the Jamaica Observer and answer the questions beneath.

For electric cars to take off, they'll need a place to charge

Around the world, support is growing for electric cars. Automakers are delivering more electric models with longer range and lower prices, such as the Chevrolet Bolt and the Tesla Model 3. China has set aggressive targets for electric vehicle sales to curb pollution; some European countries aim to be all-electric by 2040 or sooner.

Those lofty ambitions face numerous challenges, including one practical consideration for consumers: If they buy electric cars, where will they charge them?

Key terms

carbon dioxide sink removal of carbon dioxide gas from the atmosphere

reforestation planting trees to create new areas of forest

battery device for storing electricity

a) What models of electric cars are mentioned?

b) Which countries will actively promote the use of electric cars in the future?

c) What problem is foreseen in persuading people to move to electric cars?

Review of Climate change

- Global warming is the result of a small but significant increase in the average temperature of the Earth. It is thought to be responsible for such things as changes in weather patterns around the world, the melting of polar ice caps, and an increase in forest fires and droughts.

- Global warming is thought to be the result of the greenhouse effect. Some gases, including carbon dioxide, are described as greenhouse gases because, in the atmosphere, they trap heat radiation from the Earth and prevent it from passing out into space. Over the past 200 years the concentration of carbon dioxide in the atmosphere has slowly risen due to activities like the large-scale burning of fossil fuels.

- The impact of a greenhouse gas in retaining heat radiation depends on its greenhouse index and on its concentration in the atmosphere. The gases that have the highest ability to absorb heat radiation are not always the biggest contributors to the greenhouse effect.

- Weather is concerned about current conditions like temperature, rain and wind at a location while climate is about long-term conditions over a large area.

- When global warming first became an issue, some scientists believed that changes to the climate in different parts of the world were simply no more than long-term variations in conditions like temperature, rainfall, etc.

- Scientists have been accumulating data about the atmosphere and climate over a long period of time and this has allowed them to observe how the climate has dramatically changed in some parts of the world.

- Even a very small rise in mean or average surface temperature can have a huge effect on the climate of an area and its environment.

- The results of global warming are affecting all plants and animals on the planet and not just those grown and reared by people.

- Reducing climate change involves reducing the concentration of greenhouse gases in the atmosphere and particularly carbon dioxide.

- Carbon dioxide sinks, electrically powered vehicles and the increased use of alternative energy sources will reduce the atmospheric carbon dioxide level over time.

Review questions on Climate change

1. How might global warming account for the following observations?

   a) In some countries, there have been a record number of bush fires in recent years.

   b) In some areas of the oceans sea water is becoming less salty.

   c) Countries with large areas of land close to sea-level may shrink in size.

2. The current concentration of carbon dioxide in the atmosphere is 407 parts per million.

a) Express this value as a percentage.

b) Suggest why is the concentration of carbon dioxide is always measured in dry air.

c) Green plants absorb carbon dioxide during the process of photosynthesis. Would you expect to see small fluctuations in the carbon dioxide concentration between summer and winter? Explain your answer.

d) Fossil fuels like petrol and diesel release carbon dioxide when burnt. Would you expect the percentage of carbon dioxide in the air in a rural location to be the same as in an urban location? Explain your answer.

3. a) Name two gases other than carbon dioxide and CFC12 that make a significant overall contribution to the greenhouse effect

b) The relative effect of the gas CFC12 on the greenhouse effect is 25 000 times more than carbon dioxide. Explain why carbon dioxide makes an overall greater contribution.

4. The following graphs show how the average surface temperature of the land and the sea has changed since 1950 relative to a base value. The red and purple lines show the general trends.

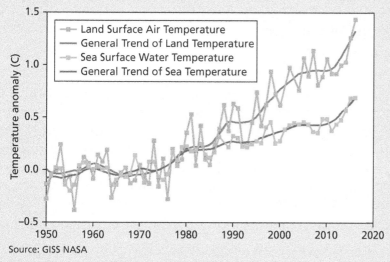

Source: GISS NASA

FIG 8.RQ.1

a) i) What trend is shown by both graphs?

ii) Is the trend equal for the surface temperature of the land and of the sea?

b) i) Was the trend less, about the same or greater between 1980–2015 compared to 1950–1980?

ii) Explain your answer to **i)**.

c) Using the smoothed values, by how much has the land surface air temperature increased between 1950 and 2017?

d) State two effects of an increase in temperature on sea water.

Climate change

· ·

We are only aware of the changes that have taken place to the Earth's climate because of the scientists who, over the years, have collected data and kept accurate records.

For example, the Mauna Loa Observatory on Hawaii has been gathering data about the composition of the atmosphere since 1956. It is thanks to their records that we can see how the concentration of atmospheric carbon dioxide has risen over the past half century.

Some schools are already involved in gathering data about weather which may one day help scientists to gain a better understanding about climatic change. Your Principal has decided that she would like your school to contribute and she has invited you to come up with a plan.

1. You are going to work in groups of 3 or 4 to investigate how your school can best obtain data about weather and keep records. The tasks are:

   - To review what sort of data would be most useful.
   - To determine what data you will collect.
   - To design, obtain or make instruments that will allow you to gather this data.
   - To determine suitable locations for setting up your instruments in order to collect data.
   - To test out your instruments at these locations.
   - To modify your instruments and review whether the locations chosen are suitable.
   - To devise methods of recording data in such a way that it will be most useful.
   - To give a presentation in which you explain what you have done and to provide recommendations which are supported by the results of your work.

   a) Look back through the unit and identify data used to describe weather and climate. For example, mean temperature and mean rainfall are important. Are you able to measure other features, such as the concentration of particles in the air that might affect air quality? Are there any organisations that you could contact for advice, such as the Meteorological Service of Jamaica?

   b) What devices can you make to gather data about the weather? You might have a look on the Internet for making simple rain gauges and wind vanes. Recording the maximum and minimum temperature requires continuous monitoring. Can devices that do this be bought cheaply? Take some pictures while making your instruments.

   c) Where can the measuring devices be set up? They need to be somewhere safe where they experience the weather but at the same time they need to be protected from extreme weather conditions, like very strong winds. Also, you don't want people or animals tampering with them. Take some pictures of likely locations.

FIG 8.SIP.1 Homemade rain gauge

d) How are you going to trial your devices and your location? How long should the trial last? How are you going to decide if the trial is a success or a failure? Your data needs to be accurate. Is there another source of data with which you can compare? For example, does the local newspaper publish information about the weather? Is any information available on the Internet? How will you decide if the location is good? For example, you might consider how easy it was to access your equipment and take measurements. Take some pictures or a video of your equipment in operation.

e) What modifications might you need to make to your equipment? You might want to make it more robust, easier to read, more accurate, simpler to use. Are there other locations you might like to try that you think will be more accessible or perhaps give more accurate data?

f) How will you record your data? A pen and a book are nice and simple but this may not be the best solution. How about using a spreadsheet? The data can be stored electronically and you will be able to carry out mathematical processes like finding means or plotting graphs more easily.

g) Prepare a presentation in which you inform people about what you set out to do and how you accomplished it. Use photographs and video to illustrate your talk and show your audience some of the devices you used. Don't be reluctant to talk about things that went wrong. Failure often provides the scientist with as much useful guidance as success.

Demonstrate how you store your data and explain the advantages of it.

Index

Note: Page numbers followed by f or t represent figures or tables respectively.

Acknowledgements

The publishers wish to thank the following for permission to reproduce photographs. Every effort has been made to trace copyright holders and to obtain their permission for the use of copyright materials. The publishers will gladly receive any information enabling them to rectify any error or omission at the first opportunity.

p6–7: Jag_cz/Shutterstock, p6: Science Photo Library/Getty Images, p7: Glow Wellness/Getty Images, p7: The Print Collector/Alamy, p7: LOUISE BARKER/AMERICAN INSTITUTE OF PHYSICS/SCIENCE PHOTO LIBRARY, p9: ZUMA Press/Alamy, p11: Voisin/Phanie/REX, p14: Nicku/Shutterstock, p14: Universal History Archive/Getty Images, p14: Iasha/Shutterstock, p14: Popperfoto/Getty Images, p14: North Wind Picture Archives/Alamy Stock Photo, p14: sciencephotos/Alamy Stock Photo, p14: Portrait Essentials/Alamy Stock Photo, p14: Pictorial Press Ltd/Alamy Stock Photo, p14: Dorling Kindersley/UIG/Science Photo Library, p16: Budimir Jevtic/Shutterstock, p18: vovan/Shutterstock, p18: gresei/Shutterstock, p18: Fat Jackey/Shutterstock, p20: wavebreakmedia/Shutterstock, p21: Science Photo Library/Getty Images, p22: Lev Kropotov/Shutterstock, p22: SpeedKingz/Shutterstock, p24: Suppakij1017/Shutterstock, p26: Matthew Cole/Shutterstock, p26: Gavs_ira/Fotolia, p26: Africa Studio/Shutterstock, p26: Horiyan/Shutterstock, p26: gloverk/Shutterstock, p26: Matthew Cole/Shutterstock, p26: Datacraft Co Ltd/Getty Images, p26: kyoshino/Getty Images, p26: Yurko Gud/Shutterstock, p26: Tim Masters/Shutterstock, p26: Photo Melon/Shutterstock, p26: litchima/Shutterstock, p26: SCIENCE PHOTO LIBRARY, p27: sciencephotos/Alamy, p27: sciencephotos/Alamy, p38: Frances Roberts/Alamy Stock Photo, p39: Blinka/Shutterstock, p39: litchima/Shutterstock, p39: Sébastien Mathier/Alamy Stock Photo, p40–41: steve estvanik/Shutterstock, p40: cowardlion/Shutterstock, p40: Wittybear/Shutterstock, p40: Phanom Nuangchomphoo/Shutterstock, p40: Westend61/Getty Images, p40: a-plus image bank/Alamy, p41: Henrik Sorensen/Getty Images, p41: XXLPhoto/Shutterstock, p41: Africa Studio/Shutterstock, p41: androver/Shutterstock, p41: Phillip Evans/Visuals Unlimited/Corbis, p41: CC STUDIO/SCIENCE PHOTO LIBRARY, p45: prill/iStockphoto, p45: Alexander Mazurkevich/Shutterstock, p45: bogdan ionescu/Shutterstock, p45: Peter Zijlstra/Shutterstock, p45: RIA NOVOSTI/Getty Images, p48: bunhill/iStockphoto, p48: t3000/iStockphoto, p48: klikk/iStockphoto, p48: VIPDesignUSA/iStockphoto, p48: Peter Gardner/Getty Images, p48: GVictoria/Shutterstock, p48: Laborant/Shutterstock, p50: Katrina Leigh/Shutterstock, p50: Pakmor/Shutterstock, p51: Carlos Caetano/Shutterstock, p51: Carlos Caetano/Shutterstock, p52: Pixmann/Alamy, p53: zkruger/Shutterstock, p53: cclickclick/Getty Images, p53: JoeyPhoto/Shutterstock, p54: Bill Boch/Getty Images, p54: Claude Huot/Shutterstock, p56: Africa Studio/Shutterstock, p58: Vitalfoto/Alamy Stock Photo, p60: Claude Nuridsany/Marie Perennou/Science Photo Library, p61: Andy Paradise/REX, p61: Voisin/Phanie/REX, p61: Jeff J Daly/Alamy, p62: Andrej/Shutterstock, p63: TTStudio/Shutterstock, p63: Thenhoi/Shutterstock, p63: Panther Media GmbH/Alamy Stock Photo, p64: Bonchan/Shutterstock, p70: Gbimages/Alamy Stock Photo, p71: Revirad Alamy Stock Photo, p72–73: Pakhnyushchy/Shutterstock, p72: Stillfx/Shutterstock, p72: Sam Edwards/Getty Images, p73: Pakhnyushchy/Shutterstock, p73: Gerd Guenther/Science Photo Library, p73: JPL-Caltech/NASA, p74: Sahani Photography/Shutterstock, p74: Tim Laman/Getty Images, p76: De Agostini Picture Library/Getty Images, p77: Image Source/Getty Images, p77: hadkhanong/Shutterstock, p78: WildPictures/Alamy, p79: Joe Petersburger/National Geographic Creative/Getty Images, p79: Cathlyn Melloan/Getty Images, p79: Cordelia Molloy/Science Photo Library, p80: Matt Jeppson/Shutterstock, p80: Morales/Getty Images, p80: The Washington Post/Getty Images, p80: M. Krofel Wildlife/Alamy, p81: Musat/iStockphoto, p81: Robert Harding Picture Library Ltd/Alamy, p82: Platteboone/iStockphoto, p82: Polarpx/Shutterstock, p82: Elenamiv/Shutterstock, p83: Diamantis Seitanidis/Dreamstime, p83: Tom Uhlman/Alamy, p83: FLPA/Alamy, p84: Visuals Unlimited/Encyclopedia/Corbis, p85: Tim Laman/Getty Images, p85: Zhukov/Shutterstock, p86:

Johann Schumacher/Getty Images, p86: AlexGreenArt/Shutterstock, p86: Arco Images GmbH/Alamy, p86: Fir Mamat/Alamy, p87: Andrew McRobb/Getty Images, p87: COLIN VARNDELL/SCIENCE PHOTO LIBRARY, p88: Glowimages/Getty Images, p88: Dennis Kunkel Microscopy/Visuals Unlimited/Corbis, p88: Biophoto Associates/SCIENCE PHOTO LIBRARY, p88: DR GOPAL MURTI/SCIENCE PHOTO LIBRARY, p89: Trinochka/Shutterstock, p89: Waj/Shutterstock, p89: OHN DURHAM/SCIENCE PHOTO LIBRARY, p89: KEVIN & BETTY COLLINS, VISUALS UNLIMITED/SCIENCE PHOTO LIBRARY, p89: SINCLAIR STAMMERS/SCIENCE PHOTO LIBRARY, p91: GEORGE MUSIL, VISUALS UNLIMITED/SCIENCE PHOTO LIBRARY, p92: Biophoto Associates/Getty Images, p92: Ed Reschke/Getty Images, p92: Dr. Gladden Willis/Visuals Unlimited/Corbis, p92: INNERSPACE IMAGING/SCIENCE PHOTO LIBRARY, p93: Heiti Paves/Shutterstock, p93: Ed Reschke/Getty Images, p93: UCSF/Getty Images, p94: BIOPHOTO ASSOCIATES/SCIENCE PHOTO LIBRARY, p96: Glow Cuisine/Getty Images, p96: DEA/P. CASTANO/Getty Images, p97: DEA/P. CASTANO/Getty Images, p98: Visuals Unlimited/Corbis, p99: FRANK FOX/SCIENCE PHOTO LIBRARY, p102: Lebendkulturen.de/Shutterstock, p102: Lebendkulturen.de/Shutterstock, p102: Rattiya Thongdumhyu/Shtterstock, p103: Carolina Biological/Visuals Unlimited/Corbis, p107: Amawasri Pakdara/Shutterstock, p107: David R. Frazier Photolibrary/Alamy, p113: momopixs/Shutterstock, p116: Dr. Gladden Willis/Visuals Unlimited/Corbis, p116: POWER AND SYRED/SCIENCE PHOTO LIBRARY, p119: Dave and Sigrun Tollerton/Alamy, p120: Ed Reschke/Getty Images, p123: Anthony Mercieca/Getty Images, p124: BIOPHOTO ASSOCIATES/SCIENCE PHOTO LIBRARY, p128–129: SCIENCE PHOTO LIBRARY/SOHO/ESA/NASA, p128: Pakhnyushchy/Shutterstock, p128: NASA, p129: Erdosain/istockphoto, p129: Erdosain/istockphoto, p129: Pindyurin Vasily/Shutterstock, p129: AdamEdwards/Shutterstock, p129: jokerpro/Shutterstock, p129: Joe Raedle/Getty Images, p129: ADRIAN DENNIS/Getty Images, p130: William Berry/Shutterstock, p130: Elena Pominova/Shutterstock, p131: Jeff Foott/Getty Images, p132: Peter de Clercq/Alamy, p132: Premaphotos/Alamy, p134: Yegor Korzh/Shutterstock, p134: Berislav Kovacevic/Shutterstock, p134: Greg Balfour Evans/Alamy, p135: Stefano Tinti/Shuttestock, p136: Adams Picture Library t/a apl/Alamy, p136: NASA, p138: Mike P Shepherd/Alamy, p138: Art Directors & TRIP/Alamy, p139: Alex Potemkin/Getty images, p139: Jamaican Public Service Company Limited, p140: MichaelUtech/iStockphoto, p140: rangizzz/Shutterstock, p142: Oleksiy Mark/Shutterstock, p142: Aleksey Klints/Shutterstock, p142: Joop Zandbergen/Shutterstock, p143: Monty Rakusen/Getty images, p143: MARK WILLIAMSON/SCIENCE PHOTO LIBRARY, p144: Laurence Gough/Shutterstock, p144: avleMarjanovic/Shutterstock, p144: wang song/Shutterstock, p144: DeAgostini/Getty Images, p145: pedrosala/Shutterstock, p145: Chukcha/Shutterstock, p146: Universal Images Group/Getty images, p146: Universal Images Group/Getty images, p146: Horizons WWP/Alamy, p147: yingphoto/Shutterstock, p147: Jason Lindsey/Alamy, p148: Konjushenko Vladimir/Shutterstock, p148: Dutourdumonde Photography/Shutterstock, p148: Kattiya.L/Shutterstock, p148: PCN Photography/Alamy Stock Photo, p149: CollinsChin/iStockphoto/, p149: Natalia Macheda/Shutterstock, p149: Ensuper/Shutterstock, p149: Andrey_Popov/Shutterstock, p149: Theerawut_SS/Shutterstock, p149: sciencephotos/Alamy, p152: chuyuss/Shutterstock, p152: Neil Mitchell/Shutterstock, p152: Content Solar - WRB Energy/WRB Enterprises, p153: Dmitry Naumov/Shutterstock, p154: Hurst Photo/Shutterstock, p154: Images of Africa Photobank/Alamy, p155: Atelier_A/Shutterstock, p155: Tomasz Makowski/Shutterstock, p156: airphoto.gr/Shutterstock, p156: HELENE VALENZUELA/Getty Images, p157: Milan Portfolio/Shutterstock, p163: Anton Watman/Shutterstock, p163: Arterra/UIG/Getty Images, p163: OCEAN POWER DELIVERY/LOOK AT SCIENCES/SCIENCE PHOTO LIBRARY, p163: Leon Werdinger/Alamy Stock Photo, p164–165: Biran A Jackson/Shutterstock, p165: Ideapix69/Shutterstock, p166: DW art/Shutterstock, p167: Cloudfoam/Shutterstock, p169: Ondrej Prosicky/Shutterstock, p170: Mamsizz/Shutterstock, p171: Ben Schonewille/Shutterstock, p171: Premaphotos/Alamy Stock Photo, p171: Kamieniak Sebastian/Shutterstock, p171: Nigel Cattlin/Alamy Stock Photo, p172: Ttiborgartner/E+/Getty Images, p172: Brian A Jackson/Shutterstock, p172: David Whitaker/Alamy Stock Photo, p172: Colouria Media/Alamy Stock Photo, p172: blickwinkel/Alamy Stock Photo, p173: kisa2014/Shutterstock, p176: Dorling Kindersley/Getty Images, p176: Tim Gainey/Alamy

Acknowledgements

Stock Photo, p177: Sarintra chimphoolsuk/Shutterstock, p177: De Agostini/G. Cigolin/Universal Images Group North America LLC/DeAgostini/Alamy Stock Photo, p178: Vilax/Shutterstock, p178: NIKS ADS/Shutterstock, p178: Jo Whitworth/Alamy Stock Photo, p181: Oriori/Shutterstock, p181: Scott Camazine/Alamy Stock Photo, p182: Twin Design/Shutterstock, p182: Science Photo Library, p183: Jason Smalley Photography/Alamy Stock Photo, p184–185: watchara/Shutterstock, p185: Rubberball Productions/Tyler Marshall/Getty Images, p185: Robert Fried/Alamy, p185: Florian Kopp/imageBROKER/Alamy Stock Photo, p186: Anneka/Shutterstock, p186: Eddie Gerald/Alamy Stock Photo, p186: Science Photo Library/Alamy Stock Photo, p188: DiversityStudio/Shutterstock, p189: BSIP/UIG/Getty Images, p190: Henk Badenhorst/Taxi/Getty Images, p190: Hongqi Zhang/Alamy Stock Photo, p191: Rutgers Preparatory School/Barcroft USA/Getty Images, p192: Sebastian Kaulitzki/Science Photo Library/Corbis, p194: Sebastian Kaulitzki/Science Photo Library/Corbis, p199: Christoph Burgstedt/Shutterstock, p200: Bob Pardue - Lifestyle/Alamy Stock Photo, p200: CHASSENET/BSIP/Alamy Stock Photo, p200: Zoonar GmbH/Alamy Stock Photo, p201: Studiomode/Alamy Stock Photo, p206: Milla Kontkanen/Alamy Stock Photo, p208: Iakov Filimonov/Shutterstock, p208: Iakov Filimonov/Shutterstock, p210: Kristoffer Tripplaar/Alamy, Zoltan Kiraly/Shutterstock, p211: Jeffrey Blackler/Alamy Stock Photo, p211: Scott Camazine/Sue Trainor/Science Photo Library, p212: Dr. David Phillips/Visuals Unlimited/Corbis, p213: D. Phillips/Science Photo Library, p214: Eric SA House - Carle/Getty Images, p214: Cavallini James/Bsip/Science Photo Library, p218: Catherine Lane/iStockphoto, p218: Stickney Design/Getty Images, p219: Hero Images Inc./Alamy Stock Photo, p220: Urbanbuzz/Shutterstock, p220: Tyler Olson/Shutterstock, p222: Radu Bercan/Shutterstock, p222: Andrey_Popov/Shutterstock, p222: STR/Stringer/Getty Images, p224: Mypokcik/Shutterstock, p224: Mira/Alamy Stock Photo, p225: Dragosh Co/Shutterstock, p229: Eye35/Alamy Stock Photo, p230: chanut iamnoy/Shutterstock, p230: Mike Goldwater/Alamy Stock Photo, p232–233: Juergen Faelchle/Shutterstock, p233: Peter J. Wilson/Shutterstock, p233: PSF Collection/Alamy Stock Photo, p234: Nouk/iStockphoto, p234: Premraj K.P/Alamy, p235: David Parsons/Istockphoto, p240: Sam Diephuis/The Image Bank/Getty Images, p240: Bernhard Staehli/Shutterstock, p241: WILLIAM O. FIELD, NSIDC, WDC/Science Photo Library, BRUCE F. MOLNIA, NSIDC, WDC/Science Photo Library, p242: idreamphoto/Shutterstock, p243: Vincent Ting/Getty Images, p243: J. Lekavicius/Shutterstock, p247: Peter Titmuss/Shuttertock.